FORAGING THE ROCKY MOUNTAINS

HELP US KEEP THIS GUIDE UP TO DATE

Every effort has been made by the author and editors to make this guide as accurate and useful as possible. However, many things can change after a guide is published—regulations change, facilities come under new management, and so forth.

We would love to hear from you concerning your experiences with this guide and how you feel it could be improved and kept up to date. While we may not be able to respond to all comments and suggestions, we'll take them to heart, and we'll also make certain to share them with the author. Please send your comments and suggestions to falconeditorial@globepequot.com or 64 S. Main St., Essex, CT 06426.

Thanks for your input!

FORAGING THE ROCKY MOUNTAINS

Finding, Identifying, and Preparing
Edible Wild Foods in the Rockies

Third Edition

Liz Brown Morgan, MA, FNTP, RWP, JD

ESSEX, CONNECTICUT

FalconGuides, an imprint of The Globe Pequot Publishing Group, Inc.
64 South Main Street
Essex, CT 06426
www.globepequot.com

Falcon and FalconGuides are registered trademarks and Make Adventure Your Story is a trademark of Globe Pequot Publishing Group, Inc.

Copyright © 2026 The Globe Pequot Publishing Group, Inc.
Photos by Liz Brown Morgan unless otherwise noted
Maps by Globe Pequot Publishing Group, Inc.

All rights reserved. No part of this book may be reproduced in any form or by any electronic or mechanical means, including information storage and retrieval systems, without written permission from the publisher, except by a reviewer who may quote passages in a review.

British Library Cataloguing in Publication Information available

Library of Congress Cataloging-in-Publication Data

Names: Morgan, Liz Brown author
Title: Foraging the Rocky Mountains : finding, identifying, and preparing edible wild foods in the Rockies / Liz Brown Morgan, MA, FNTP, RWS, JD
Other titles: Falcon guide
Description: Third edition. | Essex, Connecticut : Falcon Guides, [2026] | Series: Falcon guides | Includes index.
Identifiers: LCCN 2025018467 (print) | LCCN 2025018468 (ebook) | ISBN 9781493084517 paperback | ISBN 9781493084548 epub
Subjects: LCSH: Wild plants, Edible—West (U.S.)—Identification | Wild foods—West (U.S.) | Cooking (Wild foods)—West (U.S.) | LCGFT: Field guides | Field guides
Classification: LCC QK98.5.U6 M65 2026 (print) | LCC QK98.5.U6 (ebook) | DDC 581.6/320978—dc23/eng/20250825
LC record available at https://lccn.loc.gov/2025018467
LC ebook record available at https://lccn.loc.gov/2025018468

Printed in India

The author and The Globe Pequot Publishing Group, Inc., assume no liability for accidents happening to, or injuries sustained by, readers who engage in the activities described in this book.

This book is a work of reference. Readers should always consult an expert before using any foraged item. The authors, editors, and publisher of this work have checked with sources believed to be reliable in their efforts to confirm the accuracy and completeness of the information presented herein and that the information is in accordance with the standard practices accepted at the time of publication. However, neither the authors, editors, and publisher nor any other party involved in the creation and publication of this work warrant that the information is in every respect accurate and complete, and they are not responsible for errors or omissions or for any consequences from the application of the information in this book. In light of ongoing research and changes in clinical experience and in governmental regulations, readers are encouraged to confirm the information contained herein with additional sources. This book does not purport to be a complete presentation of all plants, and the genera, species, and cultivars discussed or pictured herein are but a small fraction of the plants found in the wild, in an urban or suburban landscape, or in a home. Given the global movement of plants, we would expect continual introduction of species having toxic properties to the regions discussed in this book. We have made every attempt to be botanically accurate, but regional variations in plant names, growing conditions, and availability may affect the accuracy of the information provided. A positive identification of an individual plant is most likely when a freshly collected part of the plant containing leaves and flowers or fruits is presented to a knowledgeable botanist or horticulturist. Poison control centers generally have relationships with the botanical community should the need for plant identification arise. We have attempted to provide accurate descriptions of plants, but there is no substitute for direct interaction with a trained botanist or horticulturist for plant identification. **In cases of exposure or ingestion, contact a poison control center (800-222-1222), a medical toxicologist, another appropriate healthcare provider, or an appropriate reference resource.**

CONTENTS

Acknowledgments xiii

Preface xv

Introduction xvii

Sustainable Harvesting Guidelines xviii

Warnings and Safety xx
    Extra Warnings: For Pregnant and Breastfeeding Women, and People
    with Medical Conditions or Taking Pharmaceutical Medications xxii
    Oxalates Warning xxii

Supplies to Bring Foraging xxiii
    Forager's Checklist xxiv

How to Identify Plants xxv
    Plant Names xxv
    Leaf Identification xxvi
    Fruit and Flower Identification xxix
    Life Cycle, Root, Stem, and Special Features Identification xxix

Where to Find Wild Edible Plants: Ranges, Habitats, and Legalities xxxi
    Ranges and Habitat xxxi
    Zones Used in This Book xxxii
    Check Local Regulations xxxiii

Start Foraging and Have Fun With It xxxiv

Poisonous Plants 1
Learn to identify these plants first so that you can avoid them.

    Baneberry (Ranunculaceae / Buttercup Family) 2
    Golden Banner (Fabaceae / Pea Family) 4
    Larkspur (Ranunculaceae / Buttercup Family) 6
    Monkshood (Ranunculaceae / Buttercup Family) 8
    Poison Hemlock (Apiaceae / Carrot Family) 11
    Water Hemlock (Apiaceae / Carrot Family) 14
    Western Poison Ivy (Anacardiaceae / Sumac Family) 16

Edible and Useful Plants 18
Enjoy edible plants with caution as you learn. Always harvest mindfully and sustainably.

Agavaceae / Century Plant Family. 20
Yucca, Narrowleaf 20

Amaranthaceae / Amaranth Family 23
Amaranth, Common. 23
Amaranth, Creeping 25

Anacardiaceae / Sumac Family. 28
Sumac, Smooth 28
Sumac, Three-Leaved / Skunkbush. 31

Apiaceae / Umbelliferae / Carrot Family 33
Marked by white or yellow umbrella-like flower clusters. Do not *mistake edibles for their similar-looking poisonous relatives.*

Caraway 33
Cow Parsnip. 36
Osha. 39
Queen Anne's Lace 43

Asclepiadaceae / Milkweed Family 46
Milkweed 46

Asparagaceae / Asparagus Family. 50
Asparagus 50

Asteraceae / Aster or Daisy Family 53
Flowering plants often with showy flowers or composite florets.

Arnica, Heart-Leaved 53
Big Sagebrush 57
Burdock 61
Chamomile 64
Chicory 67
Cutleaf Coneflower 70
Dandelion 72
Fleabane. 76
Goldenrod 79
Pearly Everlasting 82
Pineapple Weed 84
Rabbitbrush, Common 86
Sage, Fringed. 89
Sage, White 91

Salsify	94
Sunflower, Common	98
Thistle, Bull / Spear Thistle	101
Thistle, Creeping	104
Thistle, Nodding	107
Yarrow	109
Berberidaceae / Barberry Family	**112**
Oregon Grape	112
Boraginaceae / Borage Family / Forget-Me-Not Family	**116**
Bluebell, Mountain	116
Brassicaceae / Mustard Family	**118**
Bittercress, Heart-Leaved	118
Pennycress	121
Shepherd's Purse	123
Cactaceae / Cactus Family	**125**
Prickly Pear	125
Campanulaceae / Bellflower Family	**128**
Harebell	128
Caprifoliaceae / Honeysuckle Family	**130**
Elderberry	130
Chenopodiaceae / Goosefoot Family	**133**

Recognizable by leaves shaped like the foot of a duck or goose.

Lamb's-Quarter	133
Strawberry Blite	136
Crassulaceae / Stonecrop Family	**139**
Roseroot	139
Stonecrop	141
Cupressaceae / Cypress Family	**143**
Juniper, Common	143
Juniper, Rocky Mountain	146
Elaeagnaceae / Oleaster Family	**149**
Canada Buffaloberry	149
Russian Olive	151
Ephedraceae / Mormon Tea Family	**154**
Mormon Tea	154

Equisetaceae / Horsetail Family 156
Horsetail 156
Scouring Rush 159

Ericaceae / Heath Family 161
Kinnikinnick 161
Mountain Huckleberry / Blueberry 164

Fabaceae / Pea Family 167
Seeds look similar to pea pods, but beware: There are many poisonous, inedible species in this family.

Alfalfa 167
Clover, Red 170
Clover, Sweet 173
Clover, White 176
Licorice, Wild 179

Fagaceae / Beech Family 182
Gambel Oak 182

Grossulariaceae / Currant Family 185
Currant 185
Currant, Prickly 189

Lamiaceae / Mint Family 191
Giant Hyssop 191
Mint 194

Liliaceae / Lily Family 197
Nodding Onion 197

Linaceae / Flax Family 200
Western Blue Flax 200

Malvaceae / Mallow Family 202
Mallow 202

Onagraceae / Evening Primrose Family 205
Evening Primrose 205
Fireweed 208

Pinaceae / Pine Family 211
Blue Spruce 212
Douglas Fir 215
Pine, Piñon 218
Pine, Ponderosa 221

Plantaginaceae / Plantain Family 224
   Common Plantain 224

Polygonaceae / Buckwheat Family 226
   Bistort 226
   Curly Dock 229
   Sorrel, Alpine 233
   Sorrel, Sheep 235

Portulacaceae / Purslane Family 238
   Purslane 238

Rosaceae / Rose Family 240
   Apple 240
   Chokecherry 243
   Plum, American 246
   Raspberry 249
   Rose, Wild 253
   Serviceberry 257
   Strawberry 261
   Thimbleberry 264

Rubiaceae / Madder Family 267
   Northern Bedstraw 267

Scrophulariaceae / Figwort Family 270
   Great Mullein 270

Typhaceae / Cattail Family 273
   Cattail 273

Urticaceae / Nettle Family 276
   Stinging Nettle 276

Violaceae / Violet Family 280
   Violet 280

Honeybees 283
   Honeybees 283

Index 287
About the Author 293

ACKNOWLEDGMENTS

I'd like to acknowledge all those who, whether due to social upheaval, war, or current cultural food norms, have lost connection to the wild foods and old foodways of our collective ancestors. I acknowledge all those struggling with that disconnection, and the catastrophic health impacts of a toxic, processed, and malnourishing food system that now spans the globe and has replaced the wholesome real foods that our ancestors enjoyed and that our bodies are designed to thrive with. I acknowledge all those interested and seeking how to transform the food culture in their own lives and communities toward one that embraces once again the reciprocity, the shared relationship of protection and nourishment, cultivation and celebration, deliciousness and vitality between humans and the plants.

To all those who are trying desperately, trying elegantly, trying purposefully to welcome back our birthright as humans on planet earth to simply eat the foods the earth provides, to create a new or reclaimed culture of respect for the plants and natural ways, I honor and acknowledge you. To all those creating a food culture—farmers and foragers, chefs and artisans, policymakers and visionaries, foodies of all stripes—more like the wholesome, regenerative, wild one I want to live in, I honor and acknowledge you.

To all the ancestors who figured it out long ago, who danced and lived with the wild plants in ways many of us now only dream of, I mourn the lost lineages and aim to restore a more thoughtful relationship with the plants.

Food culture is passed on from generation to generation, but a proper food culture and proper relationship with the plants and fruits of the earth was stolen from us. We are now collectively restoring it as we devise more appropriate, modernized foodways that celebrate the past support system and demand a love of real food. I welcome back the plants and the wisdom of relationship with them, carried through the generations.

Foragers are becoming the ancestors the generations need to right the catastrophic swerve toward processed, mutated, and poisoned foods that are destroying the health of the people and the planet.

Not everyone has to be a forager, but foraging must be part of the conversation about food, health, and sustainability and the right and best and most enjoyable and most ecologically sound ways for humans to live on and protect and restore our own health. Foraging must also be part of the conversation regarding the ecosystems of planet earth in these times of a corporatized food system, changing climate, habitat loss, species extinction, and chronic health conditions.

I believe there is a basic, fundamental human right to hunt and fish and forage and that if government is to be relevant and useful to the people, it must protect the wild lands in enough abundance to provide for the needs of the people for abundance, sustenance, and wild wandering.

The foragers' path ahead is delicious and decadent and deeply healing for people and the planet for we are not on the path only to take and take from the wild lands. We are here to ensure the wild lands the ability to continue giving and thriving and nourishing for thousands of generations yet to come.

PREFACE

When I wrote the first edition of this guidebook in 2012, there were far fewer online and print resources about the edible wild plants of the Rocky Mountain region. It wasn't yet baked into the food culture I existed in as it is now. Back then foraging felt fringe and aspirational. Now I have friends and community and a lot more resources—books, blogs, and experts around the area—to learn from and explore with. Wild foods have become more real and more essential in my worldview and life foundations.

Imagine an entire population of hundreds of millions of humans attempting to eat without the ancestral knowledge of wild foods. New people on planet earth trying to figure out from scratch without the longstanding wisdom of the generations teaching us what they had learned and how they did it and what made people healthy. We late-1900s Americans were like aliens dropped onto a foreign planet with no knowledge about the edible plants, what to eat, or how to survive and thrive. We were an entire society sold on convenience and trying to figure out what we were meant to eat in a dominant culture that glorified fake food over real, newness over tried-and-true.

There are now many more foragers, wild-food writers, researchers, and teachers. I have my own community of friends to discuss and explore wild food with. It's now common and built in and part of my life and part of so many others' lives to weave wild foods into our regular food flow. There is a growing love and embrace and deepening wisdom around wild foods and I think we can feel it all across the Rocky Mountains.

After writing this book I became a functional nutritionist and worked with clients struggling with serious and mysterious chronic health conditions, many related to the poor-quality Standard American Diet plus stress plus toxic exposures plus other factors that impair digestion and the interconnected systems of the human body.

In doing this functional nutrition deep-healing work, I have shared and supported the journey with hundreds of people as they have tried to reclaim a more joyful and wholesome relationship with food in order to dramatically improve their health and have a more elegant experience in their bodies and lives. This recovery journey, this journey into real food, into learning how well our bodies and brains thrive with real food, is a massive undertaking with many twists, turns, and challenging barriers.

After seeing so many people struggle to simply eat the kind of wholesome food that we humans are meant to eat, and after seeing the health impacts when we don't, I came to the conclusion that the biggest barrier to good health for most people is the food culture they live in.

We shouldn't have to fight so hard to rise above processed, toxic, and malnourishing food culture norms, but to claim our health, that battle is necessary.

I envision one day living in a well-functioning food culture that gracefully nourishes the people with delicious, decadent food that is deeply healing for both people and planet.

My intention with the third edition of this book is for foraging to be a conduit to that renewed food culture that embraces and celebrates nourishing real foods, wild foods, wholesome foods that are deeply healing for people and the planet. I envision that knowledge of wild foods can become a connection between our culture's broken food norms to a widespread embrace and acceptance that we deserve better and that the hospitable earth provides better. Foraging is a way for us to step into our stewardship with the earth in a reciprocal way whereby the earth provides abundant nourishment to us and we care for its ability to live, exist, and persist.

My work as an author about foraging for wild edible foods has always been about food culture. I studied indigenous foodways as a college anthropology major. The knowledge that ancient cultures thrived with wild foods and had stable ways of hunting and fishing and foraging and cultivating food and living and eating has long been vitally important knowledge from which I better understand the world I live in and the human condition and how to create a better society that works for all people and the planet.

Wholesome food is the foundation of a happy and stable society, and knowledge about wild foods connects us to the fact that our home on earth is hospitable, that it provides for our needs, that the notion of battle and submission and scarcity have no place in the food system or in our relationship with the earth or our bodies.

My journey from the grocery store into the woods has calmed me, delighted me, nourished me, given me a sense of what it really means to be human on earth. It has allowed me to feel rooted. In a world where so many are untethered, I feel grounded. I want that relief for all those who want it for themselves. Being in a relationship with the earth as a forager is one of the best gifts of earthy existence.

To all the plant lovers and soon-to-be plant lovers, I'm happy to be on this journey with you. Have fun.

INTRODUCTION

This book is an introduction to foraging for wild edible plants in the Rocky Mountains. It will help you identify many of the common edible plants you see on the trails, hillsides, and riparian areas, and even the edible weeds that pop up in your own yard.

Most of the photos have been taken in Colorado and some in northern New Mexico. For each plant, I describe the range and habitat and likely elevations and ecosystems. I share specific descriptions of key characteristics to look for when identifying plants, and I give common look-alikes to help you narrow down what you are looking at.

I share some history and stories about plants, where they came from, who cultivated them if they were tended, and what area they are native to. I sometimes share a mention of ancient, ancestral uses and sometimes a mention of medicinal uses.

I always list which parts of the plants are edible, which parts need to be cooked, and which can be eaten raw. These lists will give you great fodder for further research as you expand your wildcrafting skills.

One great way to use this book is to choose a plant that you think you have found or that you are interested in getting to know. Then look up all the listed look-alikes. Figure out how to tell them all apart and you will be well on your way!

One of the most challenging parts of foraging is that the plant is often easiest to identify when it is not the right time to harvest. Let's say you want to harvest the leaves before it blooms or the seeds after it blooms but you only know how to identify the plant by the flowers. This can take a few seasons to remember the location and/or get familiar with the different life cycle stages of the plant before confidently harvesting at the right time of year.

Have fun exploring this book, and have even more fun out there in the ecosystems getting to know the plants. Notice them throughout their life cycle and log the different growth phases into your brain so you can remember to come back when they are at peak edibility.

SUSTAINABLE HARVESTING GUIDELINES

I'm generally a proponent of Leave No Trace (LNT) principles when in the backcountry. LNT guides us to have an imperceptible or even a positive impact on the natural areas we spend time in. The idea is to take only pictures and leave only footprints. Humans in nature, whether camping, backpacking, boating, or whatever the activity may be, should not cause negative impacts to our shared natural areas. Even small impacts add up to collectively massive ones.

There are so many of us trying to enjoy nature and restore ourselves in the same wild and natural places that our impacts do become quite huge. A campsite cleanup group in my county removed several hundred tons of trash from local campsites last year! Humans can be incredibly destructive and ruin the areas for others and can really destroy the ecosystems as well. It's vital to go into the backcountry, side country, and natural areas with mindfulness and a Leave No Trace mindset.

If I see a bit of plastic on the trail, I will pick it up and pack it out. Before leaving a campsite, I organize my crew to do a sweep for micro-trash. I don't leave anything out on the ground to blow away or be stolen by critters. Everything is stored away or tied down at all times. Certainly fires must never be left unattended. Fires must be put dead out, cold to the touch.

Generally, it is best to stay on the trails and avoid trampling the vegetation. Just imagine what would happen if all of the thousands or tens of thousands of hikers on any given trail just wandered off the trail all the time all over the hills. It would be a disaster.

All that said, foraging is not a Leave No Trace activity. It brings us to a question American environmentalists have been wrestling with for decades. Should nature be protected as a museum, to look at and observe only? Or is our relationship with nature participatory?

We set aside national parks and monuments as if they are museums or artifacts to honor. Is that all we need, is that all we should grant ourselves, or do we deserve, or require, more?

I wrestle with this too. Having written this book, which is a public call to forage, scares me a bit. Am I sending too many people into the hills to take and destroy? Will too many leave the safety of the trails and trample, dig, chop, and undo the delicate web of life that I so honor and cherish? In an era when forests are trying to recover after being mostly clearcut, where riparian areas are being drained and developed, and, like locusts, people are gobbling up every inch of beauty and ecosystem for McMansions and chemical monocropping, am I sending people to destroy the last bits of sanity we collectively share?

I've been wrestling with this concept for no less than 30 years, and the inevitable and recurring conclusion for me is that we humans require a deeper connection to and relationship with nature than the nature-as-museum concept allows. I personally would wither and shrink if I could not wander and nibble and participate with the cycles of nature as humans have always done and are meant to do. Deep relationship with nature is as essential and vitally important as breathing clean air and drinking clean water.

Some of the best, most profound, most rejuvenating moments in life are when we are out there being human on earth, as humans have been for thousands of generations. To take that away from us simply because there are 8 billion of us, simply because some people are destructive, simply because the land is being gobbled up, to take that away from us simply because we want to control the masses, is not right, is not the way to heal our broken, stressed-out, overmedicated, malnourished, disconnected civilization.

It's clear to me that more connection with nature, not less, is the balm we all need. If we are to save ourselves and create a truly workable modern civilization on earth, it must include redefining and deepening our relationship with, our love for, and our interactions with wild nature. If we are to be gardeners, tending the ecosystems of planet earth and caring for them through these dire times of mass extinction, climate change, and human population explosion, we must get to know it more deeply. We must get to love it more deeply. That relationship must become primary in our lives and in our culture and in our considerations about how to manage the earth and our human activities upon it moving forward.

Thus, I invite you to be the balm. I invite you to forage. I invite you to engage in this oldest and most wonderful of human activities. I invite you to share and celebrate wild foods. I invite you to fall in love over and over again with the hospitable, delicious, decadent earth. I invite you to be part of the movement to modernize our relationship with the earth and our reciprocal responsibility to it. I think there is no other way to leave a living planet to future generations. I invite you to change the collective conversation about where we go from here with regard to our stewardship and our duty to all the living plants and beings on this planet.

We humans must change our thinking away from the idea that we will inevitably cause harm and toward a mindset of care, participation, and love. We protect what we love, and if foraging can help many, many more of us fall in love with the wild ecosystems, I think that is a good thing.

With that, please forage gently. Take only what the ecosystems can absorb easily.

WARNINGS AND SAFETY

Since there are poisonous plants in the wild, which may look similar to edible plants, especially for beginning foragers, I recommend starting by looking at and identifying plant species before diving right into eating them. If you do not know for sure that a plant is safe to eat, you should not eat it!

Not all poisonous plants are listed in this book. Just because you don't see a certain plant listed as poisonous **does not mean it is safe** to consume. You must be sure that you have positively identified a species correctly before you eat it. There are deadly plants that can be encountered anywhere. Others may not cause death but can instead cause moderate to severe illness or lasting damage to your body. Consult many sources: books, photos from a variety of websites, and professional herbalists and botanists. Go on herb walks with experts to confirm your identifications. Until you feel you are confidently proficient at your identification skills, don't risk making mistakes.

Always remember that there is a lot you don't know. Even the experts are always learning. Pay attention to the specific uses listed. Not all parts of all plants can be eaten, and not all parts can be eaten raw.

Also, care for your friends, and create rules for yourselves that put safety first. Always pay close attention to your reactions and those of your friends to wild foods.

In some cases, anaphylaxis can result. Anaphylaxis is a very serious condition that results in the closing of the airways, hives, asthma-like symptoms, a swelling in the throat, and difficulty swallowing and breathing. It can result in death if not treated right away. **If you or anyone you are with begins to experience difficulty breathing or any of these symptoms, seek medical attention immediately.**

The basic rule of thumb when foraging is: Don't eat anything or rub it on your skin unless you are 100% sure you know what it is. The vast majority of plants will not kill you; however, a few, such as poison hemlock, certainly can. Don't take this stuff lightly. And by all means, do not ruin a lovely day in the wild by making a fatal or near-fatal mistake.

Unknown allergies or incorrect identification is always a possibility, and when you're in the backcountry, rescue may not come in time. Remoteness is wonderful, but you have to maintain focus on keeping yourself and your crew safe. It is best not to try wild edibles for the first time in the wild. Harvest them, take them home, and try small amounts in controlled settings where you have access to medical attention if needed.

It is beyond the scope of this book to give you every possible detail about identification, every detail about preparation, and even every detail regarding cautions and warnings.

As a forager, it is essential that you take it upon yourself to learn as much as you can about a plant before consuming it.

If you're curious, this book will leave you with unanswered questions. Let your questions intrigue you, and let them lead you to study more about the plants. Study with experts. Go on walks. Read more books. Research online. Look at government databases and private blogs. Use plant keys. Join online groups and discussion forums. Actively participate in the learning process.

There are many mistakes to be made in the woods, and I want you to be as prepared as possible before nibbling unknown plants. Plan to engage in your new relationship with edible plants and conduct as much research as you need to be fully, 100% ready to eat from the wild before doing so. There should be no guessing and no haphazard decisions.

You don't have to learn all the plants at once. I know it can be overwhelming! Choose one to focus on at a time. Get to really know it. Take your time. Find it in the wild. Smell it. Take photos. Read up on it. Be curious.

Finally, after you are sure you are ready, after your research has been exhausted and it all leads back to the same inevitable conclusion, then begin the process of determining whether that plant is something you want to consume.

For those who are particularly sensitive and reactive to foods, I recommend, before nibbling, conducting an allergy or sensitivity test on yourself. This is not foolproof, but it's still useful. Begin by rubbing the plant on the delicate skin of the inner forearm. Allow a few minutes to a few hours to see if your skin reacts. Look for a rash, hives, redness, itchiness, or inflammation.

Afterward, monitor yourself for three days. Notice any systemic reactions such as a swollen face or ankles, difficulty breathing, painful joints, painful muscles, hives in places other than where you rubbed the plant, headaches, dizziness, or anything else out of the ordinary.

If any of these things happen, it means your immune system is reacting to the plant, and you should not consume it. If you experience trouble breathing at any time, or any other reactions that are anything more than mild, **see a medical doctor immediately.**

Also, when consuming a new plant for the first time, it is a good idea to let someone else know what plant you are trying. Take a picture of it, and have some info ready. It can also help to have a sample on hand. If you have a bad reaction and have to call Poison Control or go to the hospital, they probably will not be familiar with the plant or any potential poisonous look-alikes. So, just as an extra precaution, be ready to explain to people what you ate and to give them plenty of info to be able to help you.

Extra Warnings: For Pregnant and Breastfeeding Women, and People with Medical Conditions or Taking Pharmaceutical Medications

Some people are in places in life where they are a bit more delicate or reactive and need to be extra careful.

Pregnant and breastfeeding mothers should take extra care. In general, I would recommend not eating any wild plants without consulting a professional herbalist or other skilled practitioners first. Many plants can cause uterine contractions, induce menstruation, and cause other potent reactions in the body. My notes in this book concerning pregnant and breastfeeding women are brief. I am not an expert in this area, and you should consult additional resources and professionals when deciding what plants are right for you and your little ones.

Those experiencing serious medical conditions, disease, or chronic illness should also be extra careful. If you are taking pharmaceutical medications, you should investigate possible cross-reactivity. Talk with your doctor and pharmacist before mixing potent wild plants with your meds.

That said, there are many wonderful wild edible and medicinal plants that can certainly help support your optimal health in these delicate times. Your body is in need of extra support and nourishment, and wild plants are powerful and exciting in this regard. It is not necessary to shy away completely! However, it is necessary to be especially aware of all the factors and to seek the advice that will help you make the right decisions.

Oxalates Warning

If you suffer from symptoms like muscle pain, joint pain, recurring candida infections, UTIs or kidney stones, exhaustion, and other oxalate-related symptoms, you might think about avoiding or decreasing foods high in oxalates. Oxalate sensitivity is generally part of a bigger picture of gut and immune imbalances, and reducing high-oxalate foods is just part of a protocol for recovery. Most people can, however, enjoy a diversity of wholesome real foods without having to worry about this issue.

SUPPLIES TO BRING FORAGING

Go out prepared with these few resources and your foraging forays will be even more fun and successful!

Camera: The best tool to bring is a camera! I am always amazed at the details I can notice when I zoom in on a picture that are not readily apparent to the naked eye. Tiny hairs in hidden places, for example, can often make the difference in positive identification. A camera is so valuable for this sort of detailed exploration. I also love zooming in on the many bugs I encounter on the edible plants. If you have questions about a plant's identification, you can take lots of photos and then go over the details at home.

Research Materials: It helps to have some research tools while in the backcountry. Bring a plant identification book or a few. You can also download regional plant apps to your phone. Refer to them frequently while analyzing what you see around you.

Storage Containers: Also bring containers. Many people use plastic bags, but I prefer to use sturdy, reusable containers. They protect delicate berries and leaves better than disposable bags without creating unnecessary plastic waste. Bring a variety of sizes, depending on what you are going to harvest. Baskets are great too, but they are best when the hiking will be easy. For longer or more strenuous hikes, containers are better because you can safely tuck them in your backpack.

Pen/Permanent Marker/Paper/Plant Journal: Also bring a pen and paper to take notes and write down questions and observations—or take notes on your device. For example, you may want to note the smell of a plant or the location you found it in. I like to write down what I think the plant is, take a picture of my notes, and then pictures of the plant. When I go through my photos later, I don't have to remember what I was thinking, where I was, or how the plant smelled because it's all right there.

Scissors/Knife: Bring a pair of scissors or a sharp knife to clip off leaves or stalks.

Trowel/Shovel: If you are digging roots or tubers, you will want a trowel or shovel. Be as delicate with the surrounding ecosystem as possible and put soil and debris back in place to cover your tracks and restore nature. Always leave a place looking as it did when you arrived.

Leather Gloves: If harvesting prickly species, bring a thick pair of leather or other gloves.

Survival Gear: It is always important to head into the mountains fully prepared. The weather can change rapidly and people do get lost, injured, and cold regularly even when they thought they were just going for a short stroll. How to survive in the mountains is beyond the scope of the book, but please always bring, at a minimum, extra water, snacks, warm clothes, wind gear, a visor, and a lighter when heading into the backcountry.

Forager's Checklist:
- __ Camera
- __ Research Materials / Books / Plant ID App
- __ Storage Containers
- __ Permanent Marker / Pen and Paper / Plant Journal
- __ Scissors / Knife
- __ Trowel / Shovel
- __ Gloves
- __ Survival Gear / Extra Layers

HOW TO IDENTIFY PLANTS

There are a lot of plants out there! So many that it can be overwhelming when you are getting started. I recommend just starting. Pick a few plants you regularly see and figure out what they are and what can be done with them. If you're feeling overwhelmed, don't worry. The Convention on Biological Diversity (CBD, www.cbd.int) says there are about 400,000 species of flowering plants worldwide, about 15,000 species of ferns, 1,000 gymnosperms (conifers), and 23,000 mosses. For comparison, the CBD says there are about 1 million species of insects, 28,000 species of fish, 10,000 species of birds, and 5,400 species of mammals (www.guardian.co.uk/science/2010/sep/19/scientists-prune-world-plant-list).

The plant species listed in this book are just a tiny sampling of all plants that you will find when foraging. There are just so many species and varieties out there. Most importantly, **do not force it to fit.** If a plant does not exactly match the description, it is not the right plant. That said, subspecies and subvarieties will look different, so you might be on the right track even if there are variations.

The scent and texture of a plant are often telling. If a species is described as being very pungent, or minty, etc., smell it to see if your senses can recognize the smell described. The bark of ponderosa pine has a strong vanilla smell. If you smell a tree and detect nothing, it is probably a different species (or it could be winter, which is when the sap is not flowing). If a stalk is supposed to be ridged, feel it, and observe it to be sure.

I recommend bringing plant ID books out in the field with you and making your identifications, as much as possible, while in the field. Take pictures, and then confirm with more sources when back at home.

Look at all factors, including leaf and flower shapes, colors, and arrangement. Consider range and habitat, size, and every detail you can think of! Learn common family characteristics.

Ready to get nerdy with taxonomy? Here are some more details about what to look for.

Plant Names

Plant names are useful in identification and in understanding the uses and other characteristics of plants. I have included a lot of common names for each plant where they exist because they are often very telling. For example, Satan's bolete is a poisonous look-alike of the king bolete. It's called "Satan" because it will make you very sick. Great mullein is also called lungwort because it is used medicinally to treat lung problems. Mallow is called cheeseweed because its tender, edible seeds look just like a tiny wheel of cheese. Yarrow is called soldier's woundwort because it can be used to stop bleeding. Fleabane repels fleas. The point is that names can also be useful in plant identification and in gaining a deeper understanding of the history and uses of the species.

The Latin names listed are as close as I could get to the currently accepted genus and species. Genus is listed first; species comes second. In some cases, it is important (because of inedible look-alikes or other confusion points) to identify the plant down to the species level. For others, like wild roses, it is not.

Leaf Identification

Though perhaps less sexy than flowers and fruits, leaves are an important feature in identifying a species. Fruits and flowers are often shorter-lived, so you are frequently left with only the leaves to evaluate. Look at the size and shape of the leaves and how they are attached to the stem. Leaf issues to consider include:

Leaf shape (oval, palmate, linear, heart shaped, etc.)
Leaf size (tiny, small, large, etc.)
Leaf arrangement (alternate, opposite, basal, whorled)
Leaf location (along the ground or up the stalk)
Leaf color and texture (darker on top, waxy, fuzzy, etc.)

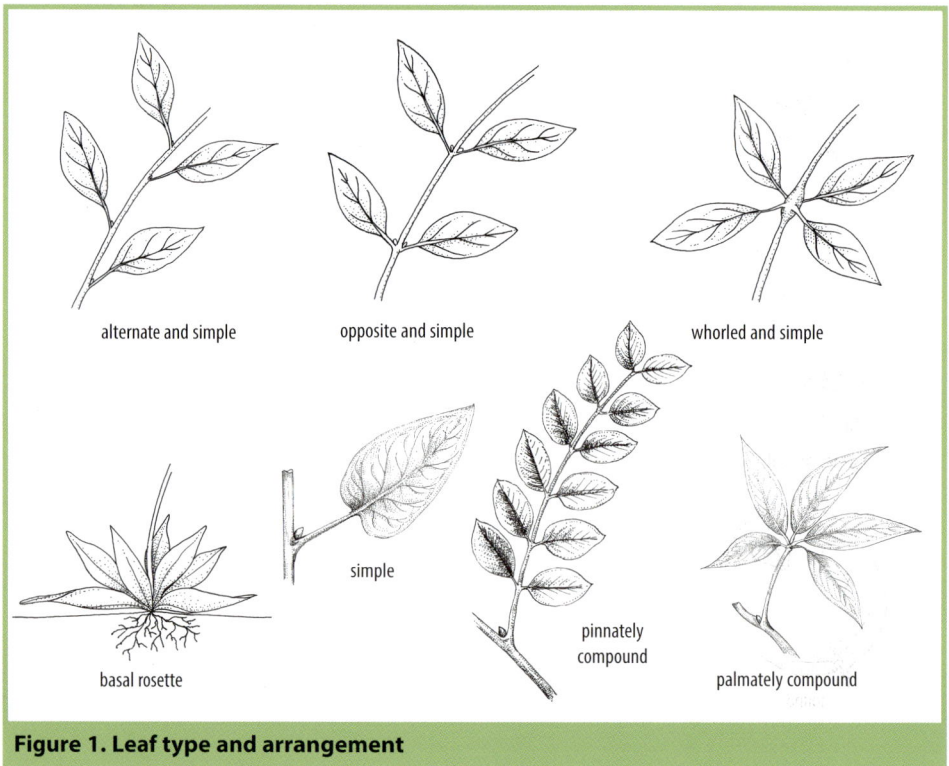

Figure 1. Leaf type and arrangement

Figure 2. Leaf margins

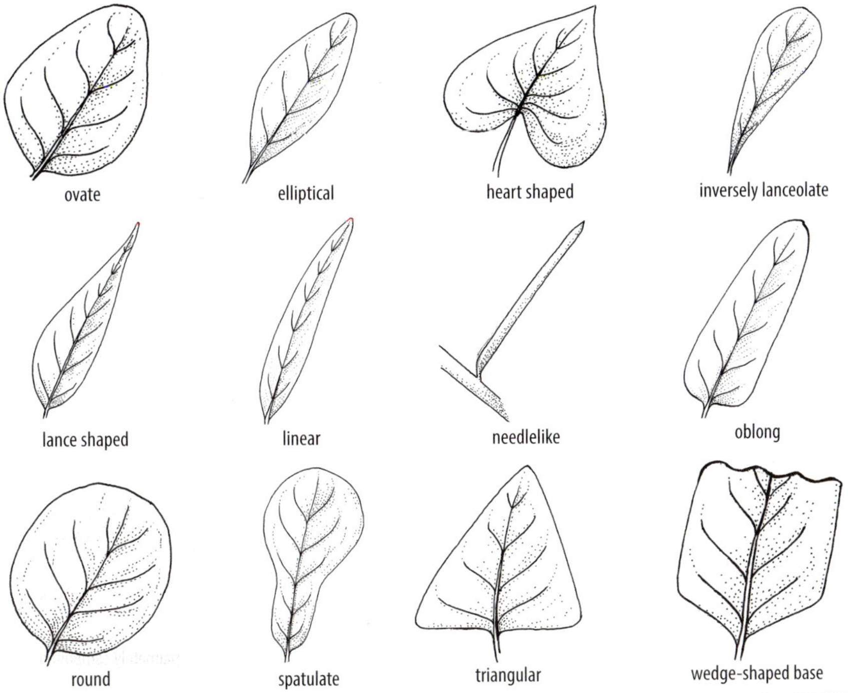

Figure 3. Leaf shapes

Basic Terms

Alternate: Leaves or leaflets arranged along the leaf stem in a staggered, not opposite, pattern.

Basal florette: A leaf cluster arising from the base of the plant. Often circular and arising in the first year for a biennial. The stalk will emerge from the basal florette in the second year.

Compound: Leaf formation where the leaf is divided into separate leaflets. The leaf blade is not continuous. Opposite of simple leaf structure. Leaflets often look like individual leaves.

Decumbent: Growth pattern indicating the plant is low-lying or grows along the ground.

Entire: Used to describe leaf margins or leaf edges that are continuous and smooth, not toothed or lobed.

Lanceolate or lance-shaped: Leaf shape that is longer than it is wide and wider in the middle than at the ends. Usually the wider portion is somewhat below the middle, tapering to a point at the leaf tip.

Linear: Leaf shape that is long and very narrow, with sides that are about parallel.

Lobed: Deep, rounded indents along the leaf edges. Can occur on any shaped leaf.

Margin: Leaf edge.

Palmate: Lobes or leaflets radiate out from a central point, usually from the top of the petiole, or leaf stem. Generally forming a somewhat circular pattern.

Pinnate: A compound leaf consisting of multiple leaflets arranged along a common axis or leaf stem. Generally forming a somewhat straight-lined pattern as opposed to palmate leaves.

Rosette: A circular cluster of leaves at the base of a plant; the basal rosette.

Serrate: Leaf margins or edges that are not smooth but toothed or jagged. Can be pointed or rounded, deeply toothed, or shallowly toothed.

Simple: A leaf that has just one part; not divided into leaflets.

Trifoliate: Clusters of three leaves or three leaflets, like a clover.

Whorl: A circular formation of three or more leaves or other structures radiating out and creating a circle or spiral around the stalk.

Fruit and Flower Identification

You will also want to inspect a plant's flowers and fruits, depending on the season. Color, shape, and how they are positioned on the stalk are important. Look at whether flowers form in a pointed or rounded cluster or as an umbel and whether they are showy or discreet. Look at petal size, number, and arrangement.

Basic Terms

Bract: A type of leaf that embraces or cups a flower or an inflorescence.

Bud: A developing, unopened flower.

Disc floret (disc flower): Small flowers in the center of a composite flower head. Lacks petals.

Floret: A small flower, typically one in a cluster making up a composite flower head. Can be ray or disc.

Inflorescence: The flowering parts of a plant; sometimes a single flower and sometimes a cluster.

Ray floret (ray flower): Small flower resembling one petal.

Umbel: An umbrella-shaped inflorescence or cluster of flowers.

Umbellet: A smaller, umbrella-shaped flower cluster that together with other umbellets make up the umbel.

Life Cycle, Root, Stem, and Special Features Identification

Also pay attention to location and habitat. You can use the range descriptions in this book to help you narrow down what plant you are looking at.

If a species is said to live only in moist, shady soil and you believe you have found it on a full-sun sand dune, this discrepancy should be taken into consideration. Could the plant have more flexibility than indicated? Sure. That's always a possibility. Also consider climate change and shifting bioregions, which will certainly continue to shift plant ranges in possibly chaotic ways now and into the future. Also consider that perhaps it simply is not the plant you think it is.

Often there will be a tiny bit of information needed to separate a particular species from another; for example, an almost unnoticeable strip of hair growing along the stalk or an obvious bloom (white coating) on the stalk.

Also look at the overall appearance of a plant. Does it grow along the ground, as a shrub or a tree? Height, width, and shape are important factors. Stem, roots, and special features should also be carefully reviewed.

Basic Terms

Biennial: A plant that lives for 2 years, often putting out leaves in year one and flowers, fruits, and seeds in the second year. Not perennial or annual.

Bloom: A white-powdery coating on a stalk or leaf.

Herbaceous: Fleshy plant, not woody; why we call herbs herbs.

Rhizome: An underground stem or root that spreads and produces new plants. Common reproduction method for perennials.

Stem: The main stalk, usually growing upward.

Taproot: Type of root that is thick and often extends deep into the earth, like a carrot.

Woolly: Having fine, soft, felt- or wool-like hairs covering the stalk or leaves of a plant, often giving a whitish or silvery hue.

WHERE TO FIND WILD EDIBLE PLANTS: RANGES, HABITATS, AND LEGALITIES

Wild edibles are absolutely everywhere. They grow on the highest mountain peaks and poke out from little cracks in sidewalks in the depths of the biggest cities. Once you start looking, you will see them everywhere: along creeks, ditches, and rivers, and growing out of canyon walls. Many plants are edible in one way or another.

You have probably walked over common plantain and red clover. You have probably sped by curly dock and wild flax on the roadside. Even many grasses are edible. Every farm in the world has abundant weeds. Almost all of these are wild edibles. Along every highway and road and trail, you will find wild edibles. If you have a garden, you can find them there too.

Wild edible plants can be native, endemic (existed in the United States before Europeans arrived), or introduced (brought here in the years after Europeans arrived on this continent). Introduced plants are also called nonnatives. Many nonnatives have become perfectly acceptable, integrated members of our local ecosystems. However, when nonnatives find themselves in habitats that are particularly well suited for them, they can become extremely aggressive and outcompete native plants, causing imbalance and often serious problems in the ecosystem. When this happens, nonnatives are considered invasive nuisance species. Eradication attempts often ensue.

The US Department of Agriculture (USDA) has an exceptional online resource for determining whether a plant grows in your area (http://plants.usda.gov). The USDA PLANTS Database provides locator maps that drill down to the county level, so you can see if a particular species has been located in a particular county. Plants migrate, and these maps are not always up-to-date, but they are a pretty excellent resource.

Ranges and Habitat

Seasons are so variable in the Rocky Mountains it is impossible to tell you exactly what time of year a wild plant will be ready for harvest. Some years, winter extends well into May with late snows and persistent cold. Other years, we have sun and warmth in January that sails right into summer.

Summer can be cold with regular hail and snow in the high peaks or in the 80s or even 90s for several months. To top it off, we have thousands of feet of elevation differentials that have a dramatic impact on seasonality and plant life cycles. If currants are ripe in one part of Colorado at one altitude, it is certain they will be weeks behind and weeks ahead in other locations.

Then there's the issue of which direction the mountainside faces! If it's facing south, it is a totally different microclimate than the north, east, or west sides. So you can see why I have very much tried to avoid anything beyond general information about when plants emerge or when they are ready to eat. You'll just have to get to know your ecosystems, and keep your eyes peeled!

The Rocky Mountain region is filled with diverse microclimates. What grows at 5,500' in elevation can be much different than what grows at 9,000'. What grows on the north (colder, darker, moister) side of a hill is usually different than what grows on the south (warmer, sunnier, drier) side. What grows in a valley is different than what grows on the adjoining hillside. What grows under a tree is different than what grows in a ditch or at the base of a warm rock cliff.

Because of these microclimates, it is difficult to give exact times of year when certain plants will sprout, flower, go to seed, and so on. In parts of the country where the terrain is more uniform, growth cycles are more consistent across the region. In our region, spring comes at least a month earlier in the lower altitude regions of Colorado than it does in the foothills nearby. Elevation is a huge factor, even though distance in miles can be quite short. Likewise, plant sizes can vary dramatically with elevation, microclimate, and moisture. The higher you go, the smaller the plants generally are. Plant height and leaf size are smaller. Fruit production might be less.

Because of this, when trying to give information about time of year to expect certain things from certain plants (when to harvest fruit, when to harvest young leaves, etc.), I have used terms such as "early spring" rather than "June," because while June might be early spring at 9,000', it is late spring or full-on summer at lower elevations. Also, seasonal variations in weather patterns, especially temperature and moisture, have a huge impact on when plants do what. Drought can cause plants to go to seed months earlier than they would in a wet year, when they grow relatively stress-free. The information I have provided should be used as a guideline, but it is up to you to assess the season and learn how plants respond in your particular microclimate.

Zones Used in This Book

Alpine zone: Also called the alpine tundra, this is the highest elevation. It is above the tree line. This biotic zone has a very short growing season, about 6 to 8 weeks. Many species of hardy plants grow here, though they all remain small as a result of ripping winds and cold temperatures. Throughout most of the US Rockies, this zone starts at an altitude of about 12,000' elevation. In the Yukon, the northern reaches of the Rocky Mountains, it can be as low as about 2,500' in elevation.

Subalpine zone: Mid- to high-range mountain elevation zone from about 9,000' or 10,000' in elevation to about 11,500' or 12,000'. Marked by

Engelmann spruce, bristlecone pine, aspen, subalpine fir, and, toward the high part of this zone, a variety of trees called krummholz stunted by the severe conditions. Also, open meadows filled with yellow glacier lily, lupine, shooting star, and columbine are included.

Montane zone: Mid- to lower-range mountainous elevations. Usually from about 8,000' to 10,000' in elevation. Ponderosa pine, juniper, Douglas fir, and Gambel oak are found in the montane zone.

Foothills: Ranges from about 4,900' to 8,000' in elevation. Species include juniper, sagebrush, piñon pine, and larkspur.

Riparian areas: Can exist in any elevation zone. Includes areas directly along water, such as creeks, rivers, flood zones, and wetlands.

Check Local Regulations

You cannot forage everywhere. Before harvesting on any public lands, check the rules and regulations. Different types of government-owned land have different rules about foraging. In some places, it is strictly prohibited. In others, it is perfectly acceptable. If you want to harvest on private lands, you will obviously need the landowner's permission.

You should also be aware that along our roads and highways, the Colorado Department of Transportation (CDOT—it may be another agency in your state) routinely sprays herbicides and poisons the plants that they consider to be noxious weeds. Some of these so-called "noxious weeds" are actually edible wild species, or they live very close to edible species. Some are covered in this book. One example is great mullein. Most counties do not have laws requiring personnel to leave signs warning of the spraying except while they are actively spraying, so you have to call the state or county to find out which areas are safe and which have been sprayed. Some counties do, however, leave signs up for a designated period after spraying. You can launch a campaign in your community for such warnings or to get them to stop doing it altogether.

I'd like to see teams of foragers work with state officials to harvest these plants. Let's see wild harvest parties replace poison trucks!

START FORAGING AND HAVE FUN WITH IT

All of the warnings aside, foraging should be a fun and relaxing experience. Don't worry about eating plants at first if you are a beginner. Just step out your back door and into your backyard or take a hike somewhere wild, and start to look at the plants. Identify them, recognize them, photograph them, and have fun. Once you have some practice and are confident with your identification skills of several species, then maybe start to forage and begin eating the bounties the earth has to offer. But don't let any of the warnings discourage you from beginning the journey. To look, smell, and photograph is harmless and is how everyone begins. So get out there and get started. Get to know this hospitable planet, this life-giving Earth, these longstanding foods, these ancient friends. Hope to see you out on the trails!

Poisonous Plants

WARNING: Do not ingest any of the species in this chapter. They are highly poisonous, and several are deadly. Please also note that this is not by any means a complete list of poisonous plants you will find in the Rocky Mountain region. In fact, some edible plants are toxic if not prepared correctly or if the wrong part of the plant is eaten. It is always important to be cautious, do lots of research, and make sure you know your plants before eating them.

BANEBERRY
Actaea rubra

Family: Ranunculaceae
Other names: Red baneberry, red cohosh, necklaceweed, snakeberry, *poison de couleuvre*, doll's eye (variation with white berries), *yerba del peco*
Look-alikes: Sweet cicely (leaves), thimbleberry (leaves), bistorts (flowers look like large bistort flowers), skunkbush (similar berries), rose hips, and anything with a red or white berry
WARNING: Very poisonous. Do not ingest. The poisonous qualities are due to protoanemonin, a poisonous essential oil found in all parts of the baneberry plant. Ingestion causes stomach cramps, burning, headache, dizziness, vomiting, bloody diarrhea, increased pulse, and circulatory problems.

Description
This native perennial has large leaves, grows in dense patches, and reaches 1'–3' tall. Stems are branched. Alternate leaves are toothed and deeply lobed. They are two to three times compound.

FORAGER NOTE: Berries are usually red but sometimes white. Often grows near edible sweet cicely; early in the season, their leaves can look especially similar.

Long flower stems shoot upward and produce rounded, elongated clusters of white flowers somewhat like a very large clover. Green seedpods form after the flowers fade away and mature into bright-red (sometimes white) berries that stand in clusters atop erect stalks flanked by dense patches of leaves.

Range and Habitat
From Alaska to New Mexico and across the northern United States. Grows up to mid-altitude, montane, and subalpine zones. Grows in moist soils and woodlands, usually in the shade.

Comments
Don't confuse baneberry with such edible berries as thimbleberry, chokecherry, pin cherry, raspberries, and currants. While there are some limited medicinal uses for baneberry root, the berries are toxic and cannot be used as food. All parts of this plant should be considered poisonous.

GOLDEN BANNER
Thermopsis spp.

Family: Fabaceae
Other names: Golden beans, false lupine, buffalo bean, golden pea, wet tooth, yellow bean, prairie buckbean, buffalo flower, yellow pea
Look-alikes: Perennial sweet pea, alfalfa, vetches, clovers, sweet clovers, wild licorice, lupine, spider flower
Related species: Spreadfruit golden banner (*T. divaricarpa*; straight or curved seedpods are glabrous, not hairy), mountain golden banner (*T. montana*; straight, erect seedpods), prairie golden banner (*T. rhombifolia*; strongly curved seedpods)

> FORAGER NOTE: The golden banners have pea pods that look similar to edible peas or string beans that you would buy in the grocery store, but don't be fooled. These are not edible peas.

WARNING: Not edible; toxic. Be extremely careful about identifications. Many members of the pea family (Fabaceae) are toxic, and many of these also have pea pods. *Do not* be tempted to eat them, even though they look similar to edible peas and beans.

Description

These common perennial members of the pea family grow up to 3' tall. The different species have similar bright-yellow, showy flowers, though they vary somewhat in seedpod characteristics, height, and the elevation where they are found.

The golden banners produce an array of showy, bright-yellow flowers that are similar in shape to other flowers in the pea family. Alternate leaves are distinctly three-lobed, although some are single and opposite.

Range and Habitat

From Washington and Montana to New Mexico. Full sun to partial shade. Moderate to high elevation. Mostly found in wooded areas and along wooded edges.

Poisonous Plants

LARKSPUR
Delphinium spp.

Family: Ranunculaceae

Look-alikes: Monkshood, wild geranium (leaves), lupine (flowers and, to some extent, leaves), harebells, mountain bluebells, buttercup (leaves), globeflower (leaves)

Related species: Dwarf larkspur (*D. nuttallianum*), Colorado larkspur (*D. alpestre*), alpine or candle larkspur (*D. elatum*), Wahatoya Creek larkspur (*D. robustum*), Sierra larkspur/tall larkspur (*D. glaucum*), tow/little larkspur (*D. bicolor*), subalpine larkspur (*D. barbeyi*), and others

WARNING: Severely toxic. Often fatal if ingested. All parts of all *Delphinium* species are severely poisonous. **Do not consume.** Symptoms can include muscular weakness, spasms, burning of the mouth, severe vomiting and diarrhea, weak pulse, respiratory paralysis, convulsions, and even death.

Description
This herbaceous native can be annual or perennial. Larkspurs have showy spikes of blue, white, or purple terminal flower clusters. Leaves are palmate with pointed tips. They are deeply lobed three to seven times. Leaves are about 6"–12" long and vary in size and shape by species. Most species have widely lobed leaves; some are thinner. Stems are erect and are 4"–7' tall.

Range and Habitat
Individual larkspur species often have limited ranges, but altogether, they reach from Alaska south through Texas and across the United States to both coasts. They can be found up to about 9,500' in elevation in meadows and forest openings.

Comments
Delphinium varieties are popular in home gardens. It is wise to visit a garden nursery and have a look at their larkspurs so that you become familiar with what they look like. There are many cultivated varieties, so keep in mind that wild varieties may look somewhat different.

Poisonous Plants

MONKSHOOD
Aconitum spp.

Family: Ranunculaceae
Other names: Wolf's bane, leopard's bane, women's bane, devil's helmet, blue rocket
Look-alikes: Larkspur, lupines, harebells, mountain bluebells, geranium (leaves), buttercup (leaves), globeflower (leaves)
Related species: Columbian monkshood (*A. columbianum*), northern/mountain monkshood (*A. delphinifolium*)
WARNING: Extremely poisonous. Do not consume in any quantity. All parts of plants contain the deadly poison aconitine. Symptoms may include paralysis of the nervous system, numbness, vomiting, diarrhea, excessive salivation, dizziness, anxiety, coma, paralysis, cardiac arrest, asphyxiation, and, frequently, death.

Description
This perennial is somewhat similar in appearance to the delphiniums (larkspur) but often shorter, growing 5" to 5' in height (depending on species). Tall,

erect, smooth stalks with deeply lobed palmate leaves growing along the stalk. Some have long, narrowly divided leaves; others have more widely divided ones.

Blue to purple or violet (sometimes white, yellow, or pink) flowers have a distinctive upper spur and bloom in midsummer. Loose clusters of three to five flowers form a raceme along the stalk. Each flower has five petal-like sepals, with the top one looking like a helmet or hood coming up and over the top of the flower.

Range and Habitat
Moist soils, along creeks, open meadows, woodlands, and wetlands up to subalpine and tundra zones. Found around the globe. Often found growing in the same habitats as larkspurs but less common.

Comments
The genus name, *Aconitum*, means "unconquerable poison." Most reports say these flowers are safe to handle, although some say numbness can occur just by touching. Monkshoods seem to be used in some cultures for medicinal purposes but only after serious and careful preparations. This is definitely *not* a plant to fool around with at home.

FORAGER NOTE: Monkshood flowers appear to have a rounded, purple hood pulled over the flower, much like that of Little Red Riding Hood. The flowers of the larkspur are somewhat pointier and sometimes more open.

Aconitum poison has been used in Europe, Asia, and the New World to tip arrows or spears with poison for hunting large animals—and for murdering humans.

Aconitum ferox, a species native to Nepal, is considered the deadliest plant in the world.

POISON HEMLOCK
Conium maculatum

Family: Apiaceae
Other names: Hemlock, devil's bread, *ciguë maculée*, *ciguë tachetée*, deadly hemlock, poison parsley
Look-alikes: All the white-umbelled species, especially osha (*Ligusticum porteri*). Also looks like osha del campo, or angelica (*Angelica grayi*), Queen Anne's lace (*Daucus carota*), wild parsleys (*Lomatium* spp.), water parsnip (*Sium suave*), and water hemlock (*Cicuta maculata*; also poisonous)
WARNING: Extremely poisonous. One of the deadliest plants on the planet. If you accidentally ingest the plant, symptoms may include salivation, nausea, coldness in the extremities, respiratory failure, heart failure, paralysis, coma, and death. If you think you have accidentally ingested this plant, contact a poison control center and get to a hospital immediately. It is always a good idea to bring a sample of what has been ingested to the hospital with you along with any photos or notes on what you think you ate.

Description
This large herbaceous biennial is a very poisonous member of the carrot family.

First-year basal leaves grow to about 18" long. Leaves are large, hairless, lance shaped, and alternate, forming lacy triangles. They are deeply but delicately toothed, creating a fernlike pattern. Leaves are pinnately divided, 12"–20" long, and 4"–12" wide. The leaf veins terminate at the tips of the serrated teeth (like osha).

The leaves are somewhat foul smelling (unlike wild carrot or osha leaves, which smell similar to carrot). But beware: The edible look-alikes are so pungent that if they are growing interspersed with poison hemlock, their overpowering scent can make everything smell similar to carrot.

The basal and lower leaves have long petioles; the upper leaves have shorter petioles. The base of each petiole is partially covered by a sheath.

In its second year, poison hemlock grows to about 3'–10' tall.

Inflorescent umbels consist of tiny white flowers atop sturdy hollow, smooth stems that stand straight and tall. They bloom from late spring to midsummer. The compound umbels are about 2½" wide. Inflorescent clusters are found from the middle to the tops of the stalks.

Stems are hollow and branching and can have purple stripes or blotches, mostly toward the bottom. The purple markings are **not necessarily a distinguishing characteristic** of poison hemlock. Many edible species of this family also have purple stems, and these markings are not always present on poison hemlock, so additional identification must be utilized. Also, a white bloom or coating that can be wiped off with your finger covers the stem of poison hemlock. Wild carrot does not have the bloom.

Range and Habitat
Naturalized throughout most of the United States, although more sparsely in the northern, far western, and far southwestern parts of the country. Found in moist and poorly drained soils, especially along creeks, ponds, and roadside ditches. Native of North Africa, Europe, and Asia.

Comments
Poison hemlock is one of the world's most famous poisons. In 399 BC it was used to execute one of the world's most famous philosophers, Socrates, who was put on trial for questioning the government and sentenced to death for doing so.

Poison hemlock contains a neurotoxin called coniine (among other poisonous components), which causes respiratory paralysis and death in humans and in many animals. I have read that artificial respiration can assist until the effects have worn off, but I sure wouldn't want to find myself in the position where this was necessary.

The plant spreads by seed and is often found in dense stands. It is considered a noxious weed in some states.

The use of the word "hemlock" can be confusing because it is used for a variety of plant species and types. Hemlock is also used to describe the *Cicuta* (water hemlock). There's also the British member of the carrot family, *Oenanthe crocata* (called water dropwort or hemlock), which is also extremely poisonous. Because they look so similar, hemlock can also refer to edible members of the carrot family, such as *Sium suave* (called water parsnip or hemlock water parsnip). Finally, there is the species *Tsuga*, a big coniferous tree also known as hemlock.

The main thing to remember is that when it comes to the group of plants with big white umbels, be very careful. Make sure you are absolutely positive of your identification before eating anything. There are several **very poisonous members** of this look-alike group.

FORAGER NOTE: The leaves of poison hemlock look much like ferns or flat-leaved parsley. Many accounts say the plant can be recognized by purple stripes or spots, but do not rely on this. The purple markings are not always present on poison hemlock, and other members of the carrot family also have purple markings.

WATER HEMLOCK
Cicuta spp.

Family: Apiaceae
Other names: Spotted water hemlock, spotted cowbane, common water hemlock, poison parsnip, spotted parsley
Look-alikes: All the white-umbelled species, including water parsnip (*Sium suave*), osha (*Ligusticum porteri*), angelica (*Angelica grayi*), Queen Anne's lace (*Daucus carota*), wild parsleys (*Lomatium* spp.), and poison hemlock (*Conium maculatum*; also poisonous)
WARNING: Extremely poisonous. Do not ingest. If you touch it, wash your hands immediately. It's probably a good idea not to smell it either.

Description
This native biennial or perennial is similar to poison hemlock, but the leaves are coarser. Leaves are arranged in pinnate clusters of three (sometimes one to two). Leaflets are sharply toothed along the leaf edges and linear or lance shaped with a pointed tip. Leaflets are up to about 4" long with a central vein and radial veins. The radial veins terminate in the notches of the leaflet teeth, not at the tips.

The root is hairless. The stalk is sometimes branched and reaches 3'–6' tall. It can have purple markings but not always.

Range and Habitat
Various species are found across the United States. Typically found in moist to wet soil, with its roots in very wet areas. Grows in high altitudes as osha does.

Comments
This is an extremely deadly plant causing severe convulsions in humans and animals. Can cause respiratory failure, extreme salivation, and vigorous chewing movements.

WESTERN POISON IVY
Toxicodendron rydbergii

Family: Anacardiaceae
Other names: Green western poison ivy, poison ivy, *Rhus rydbergii*
WARNING: Allergic contact dermatitis can occur when the skin comes in contact with poison ivy. This plant contains urushiol (as do poison sumac and poison oak), an oil that can cause a severe allergic reaction. Symptoms include hives, rash, swelling, itching, and pain. Redness, bumps, and large, pus-filled blisters can also result.

For highly sensitive people, anaphylaxis can result. Anaphylaxis is very serious and results in the closing of the airways, often with a swelling in the throat, and it can result in death. If you or anyone in your party begins to experience difficulty breathing or the throat starts to close, he or she should be taken to the hospital **immediately**.

Description
Western poison ivy is a low-growing rhizomatous perennial growing up to 4' tall. Leaves are alternate and compound, each consisting of three large, rounded leaflets. Leaflets are 1"–6" long and 1"–4" wide. They are usually shiny but otherwise can vary from fairly smooth to toothed to slightly lobed. Leaves turn from

reddish when young to glossy green in summer. In fall they turn red, orange, or yellow.

Five-petaled, nondescript, cream or light-greenish flower clusters along the leaf axils give way to white berries, or drupes. Each drupe is less than ½" in diameter and contains one seed.

Poison ivy often forms a dense ground cover and can span several square feet or even an acre.

Range and Habitat

Fertile, moist soil of all types from British Columbia throughout the United States except the southeastern states. From sea level up to 8,500' in elevation. Found along roadsides, on sand dunes, in forests, and along forest edges, especially near creeks, ditches, and areas where floods sometimes occur.

Comments

The old saying about poison ivy, "Leaflet three, let it be," is a good reminder not to touch this three-leaved plant. Related to eastern poison ivy (*Toxicodendron radicans*), which is very similar but is vine-like and climbs high into trees. Western poison ivy does not climb.

Apparently a salve or remedy against poison ivy can be made with gumweed (*Grindelia squarrosa*). I haven't tried it, but it's worth a shot if you find yourself in an itchy situation.

Edible and Useful Plants

The rest of the book is dedicated to sharing information about some of my favorite edible and otherwise useful plants of the Rocky Mountain region. Most of the plants in this section are edible, but several, while not poisonous, are not exactly what you would call deliciously edible.

I included such useful plants because some are just so incredibly common you will see them everywhere and I wanted to help you get to know the plants that you will see frequently. They also happen to be some of the most lovely, useful, and prolific plants of our region. For example, big sagebrush and fleabane are not plants you will make a meal out of, but they are still exciting, very common, and great ones to recognize and utilize in your foraging lifestyle. Arnica is another example of a plant that is not edible but has such wonderful uses and is so prolific in our region that it just had to be included!

The plants in this section are organized by taxonomic family. When you begin to understand plant classifications and common characteristics shared by members of a family, identification becomes so much easier and more fluid. When you can recognize the common leaf shape of the carrot family, or the common flower characteristics of the rose family, the square stem of the mint family, or the smell of the mustard family, you begin to create your own mental filing cabinet about plants.

When you find yourself meeting a new plant that you have not seen or identified before, this family framework will allow you to narrow down what family the plant is in more easily, and from there, identification becomes so much easier.

Please note, though, that taxonomy is a living science, and classifications do change periodically as researchers and experts learn more about the shared lineages of plants and make adjustments to more clearly reflect current thinking. You may see different species names or Latin names as you conduct further research.

AGAVACEAE / CENTURY PLANT FAMILY

YUCCA, NARROWLEAF
Yucca glauca

Family: Agavaceae
Other names: Soapwort, soapweed, Spanish bayonet, Great Plains yucca, needle palm
Look-alikes: Century plant (*Agave americana*), sotol (*Dasylirion wheeleri*), agave (*Agave*), banana yucca (*Y. baccata*)
Related species: Narrowleaf yucca (*Y. angustissima*, *Y. harrimaniae*)
WARNING: Pregnant women should not drink tea made from the roots, as it can induce labor. High saponin levels in the roots can make them toxic in large amounts.

Description
This common native perennial is a showy evergreen subshrub. It has very stiff, erect basal leaves that angle slightly outward to form a dense, rounded cluster from which an erect stem emerges and produces a white, cream, pink-tinged or green-tinged cluster of fat bell-shaped flowers. Flowers are large, bulbous, and form a tight panicle cluster along the stalk. The stalk grows 1½'–5' tall. Blooms late spring to summer.

Leaves are simple, stiff, linear, and needle- or sword-like. The very fibrous leaves grow 1'–5' long and form in clusters 2'–4' wide. Leaves shed along their margins and have fibrous strings curling along the edges. Fruits are hardened, ovular, cream or light greenish capsules about 2½" long.

Range and Habitat

Rocky areas with full sun exposure throughout dry plains and foothills from Alberta, Canada, to Texas. Hardy to USDA Zone 4. Thrives in sandy and poor soil conditions. Can be found up to about 8,500' elevation and is often seen along roadsides and disturbed areas.

Comments

The flowers, flower buds, and young stalks are edible. Flowers can be eaten raw, dried, boiled, or fried. Personally I do not enjoy them raw, though sautéed flowers are easily one of my favorite wild foods in the Rockies. Some resources recommend using only the petals and discarding what they call the bitter, green inside part. However, I find when cooked, there is only a pleasant bitterness and the entire flower is delicious to eat.

The flower stem can be used similarly to asparagus, which it resembles. I have read reports that the leaves can be eaten, but they are very fibrous and so they require long baking or boiling times. Seedpods may be edible, but I have not confirmed this.

RECIPE

Yucca Flower Sauté

I have absolutely fallen in love with a simple yucca flower sauté. Simply harvest flowers, heat up a skillet with olive oil, and sauté whole flowers until wilted and a bit browned. Add salt during or after cooking.

Another variation is to chop up the flowers, sauté in olive oil, and then add an egg. Scramble together for a yucca flower–egg scramble.

RECIPE

Yucca Shampoo

Add ½ cup yucca root (fresh or dried) to 1½ cups water. Boil until mixture becomes sudsy. Remove from heat, and use as you would any shampoo.

Variation: Can also be made using cold water.

Narrowleaf yucca is the state flower of New Mexico and is commonly referred to as soapweed because the roots can be pounded and mixed with water to make soap. The root can be made into a tea and used medicinally for inflammation and arthritis.

Narrowleaf yucca has a symbiotic pollination relationship with the night-flying yucca moth.

Its relative, the banana yucca (*Yucca baccata*), has large, soft, sweet fruits that are excellent to eat. Banana yucca is found from California to New Mexico and Texas and into Mexico, not really in the Rocky Mountains.

For harvesting flowers, wait until you find a hillside full of blooming yucca. Harvest a couple of flowers from several different stalks to spread out your impact. Please note, if you harvest the stalk, there will be no flowers that year, so harvest mindfully and leave plenty to flower.

FORAGER NOTE: *Yucca glauca*, or narrowleaf yucca, shares a similar name with an entirely different food plant. It is not the same as the starchy, potato-like root eaten widely across the Caribbean, South America, and parts of Africa, called Yuca (*Manihot esculenta*). Yuca is also called cassava, manioc, and tapioca.

AMARANTHACEAE / AMARANTH FAMILY

AMARANTH, COMMON
Amaranthus retroflexus

Family: Amaranthaceae
Other names: Pigweed, redroot pigweed, wild beet, green amaranth
WARNING: High in oxalates. Can accumulate nitrates, which can be extremely toxic. Avoid eating amaranth that has been grown in areas that have received high doses of nitrogen fertilizer.

Description
This common annual grows to about 8"–8' tall. Affected by moisture. In particularly dry or compacted soils, it will be on the shorter side.

Green flowers lack petals and form cone-shaped clusters along the stalk, especially at the top. Flower clusters can reach 11" long but will be smaller on smaller plants. They contain sharp, spinelike bracts when dry.

> ### RECIPE
> **Green Juice**
>
> Harvest leaves from a variety of wild greens such as dandelions, thistle, amaranth, wheatgrass, yarrow, or whatever you find growing. Place all leaves in a blender filled with water. Blend on high until greens are pulverized, about 2 minutes. Strain and retain the green juice. Can be stored in a glass jar with a lid in the refrigerator for a day or two. Drink one or more times per day, especially before meals.

Fruits are tiny and contain one seed each. Tiny seeds are shiny and black (sometimes dark brown). Leaves hang from petioles and are alternate, ovate (oval) with a somewhat pointed or diamond-shaped tip, and ½"–5" long.

Range and Habitat
From Alaska southward throughout the United States. Found on plains up to the montane zone, especially in gardens and other disturbed sites. There are numerous species of amaranth in the United States.

Comments
Leaves and seeds are edible and nutritious. Harvest seeds by beating them out of their pods and winnowing off the chaff. Rinse and allow to dry, and then toast lightly or eat raw. Leaves can be eaten raw, dried, or cooked. Leaves can also be dried and ground into a green powder for a nutritious addition to smoothies.

> ### RECIPE
> **Amaranth Leaf Quesadilla**
>
> Start with a large, very soft, corn tortilla. Lay tortilla out on a cutting board, and cover with a layer of cleaned, chopped amaranth leaves. Cover with cheese or cheese substitute. Add sliced tomatoes and fresh, chopped garlic. Sprinkle with fresh-ground pepper. Add second tortilla to the top.
>
> Heat 1 tablespoon olive oil in a large skillet. Place in hot skillet, and fry over medium-high heat until just brown. Flip it, and repeat until the other side is lightly browned. Remove from heat, and cut into slices. Sprinkle with salt to taste.

AMARANTH, CREEPING
Amaranthus blitoides

Family: Amaranthaceae
Other names: Prostrate pigweed, mat amaranth, matweed
Look-alikes: Somewhat like purslane and chickweed
WARNING: Can accumulate nitrogen when grown in high-nitrogen soils. Avoid eating from locations where nitrogen fertilizer is used, as it can cause problems such as blue baby syndrome.

Description
This prostrate, creeping annual reaches a height of only about 8" tall and can spread about 2' wide. Thick, succulent, red or whitish stems crisscross along the ground. Spoon-shaped leaves are cupped and ringed in white around the edges (although some varieties of this species have leaves that are not cupped and do not have the white rim). Leaves form in clusters off stem shoots and are attached by small petioles. Flowers in late summer or early fall and then sets seeds in fall.

RECIPE

Creeping Amaranth Macaroni and Cheese

Harvest 1 cup leaves of creeping amaranth by stripping leaves from the stalks. Wash well in cool water to remove dirt.

In a medium-size saucepan, combine 1 cup organic milk with 2 tablespoons butter, ½ teaspoon salt, and ½ teaspoon fresh-ground black pepper. Heat until mixture just begins to bubble; reduce heat to prevent boiling. Add the amaranth leaves, and stir for about 2 minutes. Then add 1 cup cheddar cheese and ¼ cup Gorgonzola cheese to saucepan. Allow cheese to melt, but do not boil. Stir periodically to prevent sticking.

Meanwhile, boil water and cook small shell-shaped pasta according to package instructions. Drain pasta, and add hot cheese sauce. Mix gently but thoroughly. Top with grated, chopped, garden-fresh parsley.

RECIPE

Kasha with Creeping Amaranth

Kasha reminds me of my Grandma Buddy's house. Here's a simple wild twist on an old favorite. Kasha is buckwheat and is gluten-free. Kasha has a strong, earthy flavor and is good mixed with other more mildly flavored foods. Traditionally, it is made in a dish with bow-tie pasta.

Rinse 1 cup kasha in water. Bring 2 cups water to a boil. Add the kasha, and reduce heat to a simmer. Simmer covered for 15 minutes. Remove from heat, and let stand with lid on for 10 minutes.

In a separate large bowl, add 1 cup creeping amaranth leaves. Cut the kernels off of 2 or 3 cobs of organic sweet corn. Add to bowl. Chop 1 apple into small, bite-size squares. Add to bowl. Add 2 tablespoons olive oil and the juice of 1 large lemon. Toss veggie-apple mixture so that all ingredients are coated evenly with the oil and lemon juice. Add the warm kasha; toss well. Add salt and pepper to taste. Add dried, roasted hot peppers if desired. Enjoy cold or at room temperature.

Variation: Substitute raw or grilled plums for the apple.

Range and Habitat
Found widely across the United States, especially on disturbed ground. Grows up to 8,500' in elevation, probably higher.

Comments
Leaves and stems can be eaten raw or cooked like spinach. Seeds can be eaten whole or crushed and exude a gelatinous substance when mixed into soups. The seeds of the creeping amaranth hide behind the leaves and are not terribly worth collecting except for fun.

ANACARDIACEAE / SUMAC FAMILY

SUMAC, SMOOTH
Rhus glabra

Family: Anacardiaceae

Look-alikes: Poison sumac (*Toxicodendron vernix*; **not edible**), staghorn sumac (hairy branches; edible)

WARNING: Looks similar to poison sumac, which can cause allergic reactions, irritation of the mucous membranes, and death. If inhaled, severe respiratory problems can result. Poison sumac mostly grows in the eastern United States, but you should still be careful in case it somehow manages to make it to the Rockies.

Bark, shoots, and root might be toxic. Sources are conflicting.

Description
This is a native perennial shrub or small tree with erect, cone-shaped, greenish or cream-colored flower clusters that become rust-colored or red as they fruit and turn to seed. Fruits are somewhat sticky drupes that stand erect on the ends

of smooth branches for edible varieties. Drupes are rounded but somewhat flattened.

Sumac can be low-lying ground cover or up to 9' tall. It can form thickets by spreading root systems. Hairless stems and branches are covered in a whitish, waxlike coating.

Leaves are alternate and pinnately compound. Leaflets are unlobed, sharply serrate, and oblong or lanceolate. Leaflets are arranged in opposite pairs, with one lone leaflet at the tip of each leaf stem. They form an organized and distinct pattern of eleven to thirty-one leaflets per leaf stem. They taper to a pointed end

FORAGER NOTE: Distinguish the edible species of sumac from poison sumac by the location of the flowers. In poison sumac, conical flowers or drupe clusters hang from stems that emerge from the connection point between the leaf petiole (leaf stem) and the main branch. This is the corner where the leaf grows out of the branch. Edible species have flowers and drupe clusters that stand erect directly from the tips of the branches. Poison sumac has seven to thirteen smooth-edged leaflets per leaf stem.

Edible and Useful Plants

> ### RECIPE
>
> **Sumac Juice**
>
> In a bowl, combine 1 cup sumac fruit/seeds with 2 cups cold water. Smash and pinch together with your hands to release the seeds from the dry flesh. Allow to sit for 15 minutes—longer for stronger-flavored juice. Strain through a fine colander to remove seeds. Serve at room temperature or cold, as you would lemonade.

and turn vibrantly colorful in the fall. Leaves are darker on top and lighter on the underside.

Range and Habitat
Widely found from British Columbia across the entire United States. Grows in open woods, canyons, meadows, and on dry, rocky hillsides. Largest in rich, moist soils.

Comments
Fruits, leaf, stem, and oils are edible. Most often the fruit is made into a beverage. Fruit is somewhat lemony flavored and can be eaten raw or cooked. This is a common shrub and is excellent for making juice throughout much of the year. Seeds remain erect on the branches much of the year, often to spring, and can be used year-round.

SUMAC, THREE-LEAVED / SKUNKBUSH
Rhus trilobata

Family: Anacardiaceae
Other names: Lemonade sumac, stinking sumac, ill-scented sumac, quail bush, basket bush, three-lobed sumac, lemita, polecat bush, *R. aromatica* var. *trilobata*
Look-alikes: Rocky Mountain maple, currant, gooseberry
WARNING: Some people are allergic, especially those with extra sensitivity to poison ivy.

Description
This pungent, erect to spreading native shrub grows from 2' to 12' tall and can form thickets 30' wide. Average height is around 4' tall. Arching, hairy branches form rounded thickets. Roots are branched and spreading.

> **RECIPE**
>
> **Three-Leaved Sumac Lemonade**
>
> Combine 2 cups fruits with 8 cups cold or room-temperature water. With your hands, mash the berries, separating the fruits from the seeds. Soak for 3 to 12 hours. Stir or shake several times vigorously during the soaking period. Strain to remove seeds and skin. Serve cold like lemonade.
>
> **Variation:** Add sugar or honey; blend well. Add several sprigs of mint.

Compound leaves are deeply three-lobed, often looking more like three separate leaflets that taper to points where they attach to the petiole (leaf stem). Leaves are alternate, with rounded teeth. Leaves and branches can smell similar to a skunk but not in an unpleasant way, especially when crushed.

Bright-red berrylike drupes are ¼"–½" in diameter, often with sticky hairs.

Range and Habitat

From Alberta and Saskatchewan in Canada to Colorado, Texas, and California in foothills, canyons, and dry, rocky slopes from about 2,500' to 7,500' in elevation.

Comments

Fruits and seeds can be eaten raw, cooked, or dried. Mix fruits with cornmeal to make cakes or make fruits into jam. Dried fruits often hang onto branches throughout fall and winter, making them a winter survival food for critters. Young branches can be woven into baskets and containers.

Clusters of berries dry on the branches and make a great trail snack well after summer has ended. Somewhat lemony flavored but in a smooth and refreshing way.

> **RECIPE**
>
> **Three-Leaved Sumac Tapioca Pudding**
>
> Boil 1 cup berries in 3 cups water for 15 minutes. Mash fruits and strain through a fine colander, or Foley mill, collecting the juice in a bowl below.
>
> Prepare tapioca pudding following package instructions. Use the skunkbush juice instead of water or milk.
>
> Serve warm or cold. Sprinkle a few raw berries on top for garnish.

APIACEAE / UMBELLIFERAE / CARROT FAMILY

Marked by white or yellow umbrella-like flower clusters. Leaves usually look similar to carrot tops or parsley. Do *not* mistake edibles for their similar-looking poisonous relatives. ***WARNING:*** Some members of this family are deadly. Use extreme caution.

CARAWAY
Carum spp.

Family: Apiaceae
Other names: Wild caraway, meridian fennel, Persian cumin, *C. carvi* (white flower), *C. trachypleura* (yellow flower)
Look-alikes: Poison hemlock (*Conium* spp.), water hemlock (*Cicuta* spp.); any of the white-umbelled species, including osha (*Ligusticum porteri*), osha del campo (Angelica; *Angelica grayi*), Queen Anne's lace (*Daucus carota*), wild parsleys (*Lomatium* spp.), water parsnip (*Sium suave*), sweet cicely, and yarrow
WARNING: Some sources warn that excessive use of caraway can lead to kidney and liver problems. Use in moderation as you would any seasoning. **Do not confuse with the deadly look-alikes** poison hemlock and water hemlock.

Edible and Useful Plants

Description

Wild caraway is a biennial (sometimes three-year, sometimes perennial) herb with a parsley-like basal rosette in its first year. In the second year, one or more straw-colored or purplish stalks, 1'–3' tall, arise from each taproot. Taproots are small, up to about ½" thick.

The stalk is lined with deeply divided, feathery, somewhat carrot-like alternate leaves. *C. carvi* has white or pinkish flowers that form flat-topped clusters or umbels 1½"–4½" wide at the top of each stalk. *C. trachypleura* has similar but yellow flowers.

Wild caraway fruits (achenes) look similar to store-bought caraway seeds. They are notable by their distinct, pale-colored linear ribs or ridges. They are brown, narrow, and crescent shaped.

The photo is of the less common species of wild caraway, *C. trachypleura*, which has more-rounded, less–crescent-shaped seeds than *C. carvi*.

Range and Habitat

Dry to moist, disturbed areas, pastures, roadsides, and ditches; sun or dappled shade; and plains and mountainous regions up to at least 9,000' in elevation. *C. trachypleura* is limited to a narrow band along the eastern slope of the Rockies. *C. carvi* is much more widespread.

Comments

Designated as a noxious weed in Colorado, which means it is "required to be eradicated, contained, or suppressed." Crazy. The seeds are used around the world as a flavoring in sauerkraut, pudding, curry, and liquors and as an after-meal digestive aid. Leaves are less flavorful than the seeds. Use leaves as a salad green or cooked. The root is also edible like any root vegetable and has a stronger flavor than parsnip.

RECIPE

Grilled Beets and Turnips with Wild Caraway Seed

Early fall, when the beets are huge and the caraway seeds are drying on their stalks, is the right time for some longshadowed barbecues on the back deck.

Preheat grill. Harvest garden-fresh beets and turnips, and scrub off the dirt. Slice thickly into rounds, and place in a bowl. Add some olive oil and salt. With your hands, massage oil mixture into vegetables.

Grill over medium heat, turning every few minutes. Begin cooking beets before turnips, as they take longer to cook.

Place fresh-picked parsley in bowl with remaining oil-salt mixture, and toss like a salad.

When root vegetables are just beginning to brown, remove from grill, and place directly onto the bed of parsley. Garnish with plenty of wild caraway seeds.

COW PARSNIP
Heracleum maximum

Family: Apiaceae
Other names: Indian celery, wild celery, pushki, *H. lanatum*, *H. sphondylium*
Look-alikes: Poison hemlock (*Conium maculatum*), water hemlock (*Cicuta maculata*), osha (*Ligusticum porteri*), osha del campo (Angelica; *Angelica grayi*), Queen Anne's lace (*Daucus carota*), wild parsleys (*Lomatium* spp.), water parsnip (*Sium suave*), devil's club (leaves), baneberry (leaves)
WARNING: Do not confuse with water hemlock, poison hemlock, or baneberry, which are DEADLY poisonous.

Be careful when harvesting, as the sap of cow parsnip on the skin can cause serious reactions and blistering when exposed to sunlight. This is called phytophotodermatitis and looks like dark splotches on the skin. Covering the skin with long sleeves and gloves before harvesting is recommended.

FORAGER NOTE: Its huge leaves distinguish cow parsnip from other carrot family members.

Description

This tall, robust native perennial grows from 3½' to 8' tall, sometimes taller. Large, flat-topped (or somewhat rounded) white umbels of flowers are pungent and grow 8"–12" wide. The large, rounded, maple leaf–shaped (palmate but not deeply lobed) leaves are 8"–20" wide. Egg-shaped fruit is less than 1" long.

This is one of the easier to identify of the white-umbelled look-alikes, as the leaves are uniquely palmate (not fernlike or parsley-like) and are huge compared with other carrot family relatives.

Range and Habitat

Moist or wet soil from Alaska to New Mexico in plains, foothills, forests, subalpine and montane regions, and especially in riparian areas. Up to about 10,000' in elevation.

> ### RECIPE
>
> **Ants on a Log**
>
> This is a wild twist on an old favorite and a great trail snack. Peel raw stalks, and spread peanut butter along the stem. Press plump raisins onto the peanut butter.

Edible and Useful Plants

> ### RECIPE
>
> **Sautéed Stalks and Leaf Stems**
>
> Harvest 3 cups young stalks and leaf stems. Peel and slice into 3"–5" sections.
>
> Heat a large skillet to medium-high heat. Add 2 tablespoons olive oil. Add 3 cloves fresh garlic, chopped, and gently sauté for 3 minutes. Add cow parsnip stalks and petioles. Sauté until soft, about 8 minutes, stirring frequently. Remove from heat. Add a dash of salt and crushed, dried roasted hot peppers.

Comments

Young, hollow stems can be peeled and eaten raw, roasted (unpeeled), boiled, steamed, or dipped in sugar. They taste similar to celery and also somewhat similar to rhubarb but are not good when older. The roots are also edible and taste similar to rutabagas or parsnip. Roots can be made into tea or a poultice. Young leaves are edible especially unfurled. Roots can be boiled to extract the sugar. Insides of stems can be eaten raw or cooked.

Good forage for elk, mule deer, and small mammals.

OSHA
Ligusticum porteri

Family: Apiaceae

Other names: Bear root, Porter's lovage, chuchupate, osha of the mountains, coughroot, Colorado coughroot

Look-alikes: Poison hemlock (*Conium maculatum*), water hemlock (*Cicuta maculata*), osha del campo (Angelica; *Angelica grayi*), Queen Anne's lace (*Daucus carota*), wild parsleys (*Lomatium* spp.), water parsnip (*Sium suave*), Gardner's yampah (*Perideridia gairdneri*)

Also similar to Canby's licorice-root, or Canby's lovage (*Ligusticum canbyi*). Also sometimes called osha, which is found only in the northwestern United States and British Columbia.

WARNING: Do not confuse with deadly look-alikes poison hemlock and water hemlock. The root of osha is strong and medicinal. Use with care. The root might be too strong for use during pregnancy.

Description

This tall, pungent native perennial has erect, hollow stems with white, flat-topped flower heads.

Large upright or upward-arching fern- or parsley-like leaves grow along the stalk. They are larger and denser, up to 2'–3' high, becoming much smaller and sparser on the upper half of the stalk. The largest leaves can reach up to about 2' long. Leaves are alternate, with a strong carroty or parsley scent. Young leaves emerge in a tight, ruffled cluster and, eventually, unfurl. Leaves are green but turn yellow, orange, or reddish in fall. Leaf veins terminate at the tips or points of the teeth (like poison hemlock), not in the margins or dips between the points (like water hemlock).

A hollow, erect stem emerges from the cluster of basal leaves and grows 2'–6' tall. The stem is green and can have purple splotches or markings. The root is brown and has obvious remnants of past years' growth, which forms a fibrous, thick, peeling, hairlike skin around the fresh root.

Flowers in late summer. Flower heads emerge as tight whitish umbels and unfurl to become wider, white, flat-topped compound umbels.

One way to distinguish this plant from its poisonous relatives is by its strong carrot smell. However, be aware that it can grow interspersed with poison hemlock or the poisonous water hemlock, and its smell can overwhelm the area, making it difficult to tell which plants do not share the scent. Be sure to use ALL other factors in making an identification.

Range and Habitat

Grows from about 5,000'–10,000' in elevation. It is found in moist, high-altitude areas, especially with some dappled shade, mostly in the Four Corners region. Often grows in dense patches under aspen groves. More sparsely found in Montana, Idaho, Wyoming, and Nevada.

Many sources claim that osha and poison hemlock grow at different altitudes; however, this is not the case and can't be relied on for identification.

Comments

Osha is a plant only for advanced foragers. It is listed here only so that readers can be aware of this wonderful and well-loved herb. It so closely resembles several

> RECIPE
>
> **Raw Honey and Osha Root (For Colds and Flu)**
>
> Chop dried osha root into small pieces or slivers. Fill a jar halfway with raw honey—locally sourced if possible. Add chopped osha and stir together with the honey. Suck on honey-soaked osha at the first sign of a cold, flu, or respiratory distress.

Edible and Useful Plants

> **RECIPE**
>
> **Osha Root Tea**
>
> Bring water to a boil. Place osha root pieces into tea strainer and pour boiling water over it. Steep for 7–10 minutes. Drink slowly at the first onset of a cold, flu, or sore throat. Excellent with a spoonful of honey and dried peppermint leaves.

deadly poisonous plants that you should absolutely not plan on consuming it unless you have been trained by someone with significant expertise. This is not something like dandelion or raspberry, or even the similar-looking cow parsnip, that you can be confident in identifying with just the advice of books and some careful investigation.

Another issue is that osha is such a beloved medicinal and nearly impossible to cultivate, that in many areas it has been overharvested. Harvest small amounts only and from large stands.

I do, however, recommend searching for it and using that search to hone your plant identification skills. Thinking about the significant consequences of an incorrect ID will get your heart pumping and will keep your senses alert as you go through each piece of the identification puzzle.

Leaves, roots, and seeds are edible. Leaves can be eaten raw, cooked, or dried. They can be considered medicinal but are mildly so and can be eaten like any green. The seeds have a spicy, celery-like flavor and are good to use as a seasoning cooked or raw. The roots are very strong and are medicinal but not food. Root is used dried as a tea, decoction, or tincture for coughs, congestion, and other medicinal purposes.

I have read that a traditional native food use of this plant was to pick osha leaves prior to flowering. They were then dried and used in soups and stews.

QUEEN ANNE'S LACE
Daucus carota

Family: Apiaceae
Other names: Wild carrot, bird's nest (because the old flower curls into a bird's-nest shape), bishop's lace
Look-alikes: Poison hemlock (*Conium maculatum*), water hemlock (*Cicuta maculata*), osha (*Ligusticum porteri*), osha del campo (Angelica; *Angelica grayi*), wild parsleys (*Lomatium* spp.), water parsnip (*Sium suave*), yarrow
WARNING: Very easy to confuse with deadly look-alikes poison hemlock and water hemlock. Foragers in our region have died because of this confusion.

> FORAGER NOTE: One of the easiest of the white-umbelled look-alikes to identify because of the tiny red or dark-purple flower in the center of the umbel—which lore tells us is a drop of blood spilled in the center of the umbel as Queen Anne pricked her finger while sewing her lace. The "blood" droplet is actually a single, tiny, dark-colored flower right in the middle of the white umbel.

Edible and Useful Plants

Description

This biennial creates a first-year basal rosette that looks similar to the leaves of a cultivated carrot. The parsley- or fern-like leaves are low growing, although, with competition, they stand more upright. In the second year, the stalk reaches 2'–6' tall. The stalk is sometimes branched, especially toward the top. The stem can be distinguished from that of poison hemlock because it is solid (not hollow) and does not have a white coating (bloom). Hairy stem is light green with purple stripes.

When flowering, the plant is typically shorter than poison hemlock, which more commonly grows 6'–9' tall. Leaves and stem of wild carrot are noticeably hairy, whereas poison hemlock is hairless.

Umbels form mostly at or toward the top of the plant (rather than from the middle upward, like poison hemlock), although stands often have shorter and taller plants, so there is an illusion of flowers growing from the middle. Flower clusters consist of about thirty umbellets that form the umbel, which is about 5" across. The center umbellet often contains a single purple flower.

Far thinner and more delicate looking than the similar-looking Angelica species.

Range and Habitat

Grows very sparsely throughout the Rocky Mountain region. Keep this in mind when you think you have found wild carrot. It is more likely to be one of the deadly poisonous look-alikes. Much more common in the Midwest and the

> ### RECIPE
>
> **Hearty Wild Game Stew**
>
> In a slow cooker, combine 3 cups bite-size squares of wild game of choice (or beef or chicken), 2 to 4 cups thickly sliced wild carrot, thickly sliced celery (3 stalks), and ¼ cup dried barley. Add plenty of fresh-ground black pepper and sea salt, ¼ teaspoon turmeric, and several fresh, peeled garlic cloves. I also like to add a jalapeño or other spicy pepper. Bring to a boil, then simmer on low heat for about 8 to 10 hours.
>
> **Variation:** Add 1 tablespoon of honey or molasses.

eastern United States. Considered a noxious weed in the Midwest, even though the plant is edible and provides an important food source for pollinators. Be sure to find out whether stands are sprayed with herbicide by your local authorities before consuming. Grows in open fields, dry areas, ditches, and disturbed sites. Prefers the lower elevations of the plains regions throughout the Rockies.

Comments

The root, young shoots, and seeds are edible. The root is white or cream-colored and smells similar to a carrot. Use it as you would a domestic carrot.

Daucus carota is the ancestor of the domesticated carrot, and relatives are known to have been cultivated in Iran in the tenth century and in Northern Africa and Syria by the eleventh century. Many European paintings from the 1600s include depictions of carrots. This is a widespread edible food and has been for thousands of years.

Bring a shovel to dig the roots. Be sure to harvest the correct root.

> ### RECIPE
>
> **Carrot and Parsley Salad**
>
> Finely chop equal amounts young wild carrot and parsley shoots. Toss well with olive oil, sweet rice vinegar, and a dash of soy sauce.

ASCLEPIADACEAE / MILKWEED FAMILY

MILKWEED
Asclepias speciosa

Family: Asclepiadaceae
Other names: Showy milkweed

Look-alikes: Dogbane, false hellebore (*Veratrum* spp.; **poisonous**)
Related species: Common milkweed (*A. syriaca*), found in the Midwest and East, with some range overlap
WARNING: Do not consume raw sap. White milky sap in all parts of the plant is toxic raw but perfectly safe when cooked. **Many sources warn not to consume brown seeds.** Seeds and seedpods are edible (cooked) when young, but older brown seeds are reported to be toxic. Avoid getting sap in your eyes.

Several species of false hellebore (*Veratrum* spp.) have the potential to be confused with showy milkweed. Both have thick, erect stalks; grow 2'–4' tall; and have very large, thick leaves. Hellebore leaves are deeply ridged and look somewhat like big fat pieces of lettuce that would make a great burrito wrap. Don't be fooled. **False hellebore is poisonous and should not be consumed.**

Description

This native perennial stands on a thick, erect, hairy stalk 2'–5' tall. Large ovate leaves on short petioles are opposite and have a pronounced light pinkish or cream-colored midvein. Leaves grow up to 8" long. Plant reaches 2'–3' tall.

Terminal clusters of showy, star-shaped, pink flowers bloom in midsummer. Flower hoods (arms of the star) are long and pointed. Compare with common milkweed, which has short hoods. A large, distinct seedpod or follicle points upward. It is bumpy, like it is covered in warts, and contains many flat, tightly packed seeds attached to shiny, white, silky plumes.

Range and Habitat
Native to the western half of the United States. Found from British Columbia to Texas, from sea level to 9,000' in elevation. Grows in areas with some moisture, such as roadside ditches, moist fields, or other poorly drained areas.

Comments
Milkweed is especially important to the monarch butterfly, as it's one of the very few plants monarchs will lay their eggs on. Eggs are laid on the underside of dense patches of leaves, and the larvae will eventually build their chrysalis on the milkweed. Milkweed sap contains chemicals that are thought to make monarchs unappealing to would-be predators.

Monarch butterfly populations are in massive decline due to habitat destruction, including but not limited to Midwestern farmers destroying milkweed habitat and poisoning fields near remaining stands. Plant milkweed in your organic garden to attract monarchs and help save this iconic species from extinction.

For this reason, I cannot recommend harvesting wild milkweed. However, it is a fun plant to know about and plant in your garden or elsewhere.

Young shoots, stems, flower buds, flowers, and young leaves can be eaten. Many recipes suggest boiling. Very young seedpods (seeds must be white) and roots can also be eaten; again, these are usually boiled. Oil can also be consumed. Milkweed vegetables have a taste somewhat similar to peas, tomatillos, or very mild roasted green chilies.

Be sure to cook well. Milky sap (latex) should not be eaten raw, although some people eat very small amounts raw without obvious signs of problems. Safe when cooked thoroughly.

Several sources warn that the brown seeds are toxic and should not be eaten.

RECIPE

Stir-Fry with Milkweed Flowers

Harvest young flowers, which are somewhat similar to broccolini heads.

Heat olive oil in a wok over high heat. Add garlic and sizzle. Add diced milkweed flower heads. (I like to leave them whole so that you get the full visual of what the plant looks like.) Sauté until tender.

Serve over sticky rice with soy sauce and diced green onions.

Many people like to harvest while wearing gloves to keep the milky sap off of their skin and clothes. Old dried stalks can be made into cord, rope, netting, and coarse fabric.

Attracts butterflies, bees, and a variety of other insects.

The silky down of the seedpods is said to be several times more buoyant than cork and can be used to stuff pillows and cushions. It is also said to be used to stuff life jackets or personal flotation devices (PFDs), but I definitely would not recommend it in any serious whitewater—but perhaps it can be used as a fun experiment on a calm lake.

The root is used medicinally.

ASPARAGACEAE / ASPARAGUS FAMILY

ASPARAGUS
Asparagus officinalis

Family: Asparagaceae (formerly Liliaceae)
Other names: Garden asparagus, sparrow grass
Related species: *Asparagus officinalis* subsp. (Dumort.)
Look-alikes: Somewhat like the young shoots of horsetail (*Equisetum arvense*) and less like scouring rush (*Equisetum hyemale*)

Description
This erect, herbaceous perennial is edible when asparagus stalks emerge from soil in early spring. Look for asparagus stalks sticking straight out of the ground with no further pomp or circumstance until they begin to leaf into fernlike or miniature Christmas tree–like flower stalks later in the season. Look for the round, red balls (fruits) hanging from the stems, similar to ornaments. Do not eat these red berries.

Very early shoots are often purple and change to green as they grow. Wild asparagus looks exactly like the cultivated asparagus you are used to. I have seen reports proclaiming asparagus up to 8' tall; though wild in the Rocky Mountains, it is about 1'–2½' high. In the wild, height and thickness vary depending on growing conditions.

Asparagus propagates mostly through the root systems but sometimes by seed. If you come across a stand of asparagus with seeds/fruits hanging from the branches, you can poke some holes in the soil and plant a few seeds.

Range and Habitat

Native to temperate and Mediterranean zones of the Old World. Known to have been used in parts of Europe, Egypt, and western Asia for at least 2,000 to 3,000 years. This species was introduced to the United States and is now found throughout the lower 48 states, Canada, and Alaska.

Found in wet and moist habitats, next to streams, ditches, seashores, and along riverbanks. Grows in fertile and sandy soils. Often found near old homesteads, old abandoned mining communities, and in moist areas where early settlers landed and made a go of it throughout this continent, or downstream from such settlements.

RECIPE

Grilled Asparagus

Nothing says springtime like BBQ and fresh asparagus! Rub asparagus stalks gently with olive oil. Sprinkle sea salt and pepper over them. Grill over hot coals about 3 to 4 minutes per side until just able to see brown streaks from the grill grate. Turn once and eat. Grill times vary depending on how hot your coals or how large your flame, so adjust accordingly.

Excellent with a medley of fresh grilled veggies, or add grilled asparagus to the top of a fresh spring salad. Also excellent as a side dish along with grilled meat or fish or in an omelet.

Asparagus does not thrive at very high altitude, though I have seen smaller stands at around 8,000'. In general, it will be more productive to search near waterways below 7,000'.

Comments

Spring is wild asparagus season! Search along streams and riverbanks, irrigation ditches, ponds, and lakes. Never harvest an entire stand. Harvest responsibly, and leave plenty to feed the roots and go to seed to keep the stand strong for next year.

Stalks can be eaten raw or cooked. Fresh, wild asparagus will be far crisper and more tender than those bought in a store. It is so tender it does not need cooking, but, of course, there are many wonderful ways to cook asparagus too. I recommend a short cooking time, whatever method you choose.

ASTERACEAE / ASTER OR DAISY FAMILY

Flowering plants often with showy flowers and composite florets.

ARNICA, HEART-LEAVED
Arnica cordifolia

Family: Asteraceae
Other names: Leopard's bane, heartleaf arnica
Look-alikes: This is a member of the large group of yellow aster look-alikes, both lovingly and disdainfully referred to as the Damn Yellow Composites (DYCs). There are many yellow-flowered asters that can be confused, such as sunflower, dandelion, mules ear, and balsamroot.

Related species: Leafy arnica (*A. chamissonis*) grows from 8" to more than 3' tall and has 5–10 pairs of leaves. It prefers moister habitat.

WARNING: Poisonous—not edible. Arnica is not typically taken internally except under the close supervision of a skilled herbalist. It can cause severe blistering of the throat, digestive tract, internal organs, and skin. **Use caution.** Do not use salves made from this plant on open wounds.

Description

This herbaceous, native perennial is a common species in the Rockies and grows about 4"–2' tall.

Heart-leaved arnica is named for the long, heart-shaped basal leaves. Leaves along the stalk can be lance shaped. Leaves are opposite along the stalk but can become alternate toward the top of the stalk just below the flower.

Flowers are typical of the DYCs. They are bright yellow and somewhat resemble wild sunflowers but are smaller and more delicate.

The plant reproduces both by seed and spreading roots. It flowers sporadically from different root nodes from June through August.

Other arnica species flower from multi-branching herbaceous stems.

Range and Habitat

Found from the Yukon to New Mexico, from the foothills up to subalpine slopes. Heart-leaved arnica is common and grows in moist shade or dappled shade, especially in pine and aspen groves. Found from about 4,000' in the more northern regions (beginning higher in southern region) up to timberline. I see it along many forested trails throughout Colorado.

Comments

Arnica is known for soothing sore muscles and is often used as a salve to help heal injuries, bruises, and calm painful muscles.

One herbalist I consulted thought arnica should be included in the poisonous plants section rather than in the useful plants section.

I have included it here because it is so loved as a salve for sore muscles and because it is a well-known species, and is widely purchased as an ingredient in commercial salves and lotions for sore muscles and arthritis. I wanted to introduce readers to the fact that arnica is a common wild plant in our region, but it is not to be considered edible.

RECIPE

Arnica Oil

This is very potent. Use with care.

Harvest basal leaves of arnica (can also use flowers if available). Dry in a cool, dark place on screen racks. When leaves are crisp and dry, place in blender or clean coffee grinder and pulverize into a powder. You can also crush by hand or with a mortar and pestle.

Combine ¾ cup dried arnica leaves with 2 cups good oil, such as organic olive, jojoba, or sesame oil. Close lid, and store in a dark place like a cupboard or closet. Each day, gently jostle or turn the jar upside down. Let sit for a month or so.

After 1 month, filter the arnica out with a fine sieve, colander, or cheesecloth. Rub the oil on arthritic joints, areas of chronic inflammation, or areas of acute injuries, such as a sprain or bruise.

NOTE: Plant strength will vary. Some salves will be much stronger than others. Try a small amount on your skin to make sure it does not cause blisters. If it causes blisters or skin irritation, dilute with additional oil.

All parts of the plant can be used for salves, but salves should be used only on unbroken skin. Arnica dilates blood vessels and helps reduce swelling.

Arnica is typically one of the early pioneers to reemerge after forest fires. **Not edible. Not for internal use. Use extreme caution. External use also can be hazardous.**

BIG SAGEBRUSH
Artemisia tridentata

Family: Asteraceae
Other names: Great basin sagebrush, *chamiso hediondo*
WARNING: Some people have negative reactions when this plant is rubbed on skin or taken internally. **Should NOT BE TAKEN in any form (internally or externally) by pregnant women,** as it can stimulate menstrual flows.

Description
Six subspecies are recognized, including basin big sagebrush (the largest), Wyoming big sagebrush, and mountain big sagebrush. Depending on the subspecies, mature plants are often 3'–4' high, but in some cases they can be 8'–13' tall. Leaf tips are generally lobed (wavy toothed) but are sometimes pointed.

In general, big sagebrush has a woody main stem with vegetative branches reaching upward, creating an even or uneven top, depending on the subspecies.

Leaves are silvery sage–colored and deliciously pungent. They are small, less than 1" long, flat, and wedge shaped, often with three teeth at the flat tops, tapering to the base where they attach to the stem.

Edible and Useful Plants 57

Small, discreet, yellow flower heads about ⅛" long form together into narrow clusters that stick out from the crown. The clusters are stunted in drought years but otherwise somewhat tall. Achenes (small fruits) are usually smooth (glabrous) but can be hairy, though that is less common.

Range and Habitat
Found from British Columbia across the West to New Mexico and North Dakota to Nebraska. Dominant shrub of the Great Basin region. Valley bottoms, low foothills, and large expanses of dry, open plains and hillsides up to the montane zone, especially where rainfall is 12"–18" per year.

Comments
In the windy and sparse high deserts, you would feel alone if not for the big sagebrush welcoming you on every gust of wind. This is the aromatherapy of our region, the historical scent of the West. But don't confuse it with (or substitute it for) culinary sage. Wild sage has a much stronger flavor, although it can certainly be used very sparsely in cooking. The culinary sages, the Salvias, have square stems and are related to mint.

RECIPE

Foot Soak

If you have sore feet, you'll love this sage foot soak.

Harvest leaves of several branches of big sagebrush. These can be used right away or dried for later use. Bring a large pot of water to a boil, and then remove from heat. Add 1 cup big sagebrush leaves stripped from stems. Add a scoop of Epsom or other bathing salts. Stir. Place lid firmly on pot.

Let steep until just the right temperature to soak feet in. Pour hot water into a tub or basin large enough for both feet to soak. Soak feet together in basin.

Sit back, relax, and enjoy. After 5 to 10 minutes, remove one foot at a time and give it a good massage and place back into the warm bath. When water has cooled, remove feet and gently dry with a soft towel. Cover your feet in oil and put on thick, soft socks made of natural fibers. Curl up with a good book or a hot cup of tea.

RECIPE

Ceremonial Burning

Harvest sagebrush leaves and branches. Dry in a dark place with sufficient airflow. To celebrate the change of seasons, fall equinox, or winter solstice, build a small fire outdoors. Allow the wood to form embers. At sunset or sunrise, sprinkle dried sage onto the embers or place a branch over coals. Gently use your cupped hands to wash the smoke over you. Especially notice the moments when the smoke goes from being visible to the eye to becoming invisible. Allow this transition to remind you that not all things in life are always readily apparent to us. Remember that this shrub has been used in human ceremonies for thousands of generations.

I like to rub fresh leaves on my arms and neck as a sort of perfume. I also swear it keeps the mosquitos away although this is completely unscientific and unconfirmed.

Big sagebrush smells very similar to the smaller sages that are in the *Artemisia* genus. They can all be made into smudges for ceremonial burning and cleansing practices.

Leaves and seeds can be eaten raw or cooked or made into tea. Seeds can be dried and crushed and used as a flavoring.

The plant has lots of medicinal uses: as a chest compress to break up mucus; a tea to induce sweating and help with viral infections; and an inhalant when sick. Steep in oil and apply to stiff joints (good for arthritis), or make a footbath and soak. Also good to simmer in hot water and inhale aromatics when you have a cold. High in volatile oils, so steep tea with lid on to preserve the oils. Sagebrush tea is very strong; only very small amounts of weak tea should be consumed if at all.

CAUTION: Gagging can result if overconsumed. **NOT FOR PREGNANT WOMEN.**

Sagebrush is also a disinfectant and can be used as a cleaning agent in the home, and used as a tea poured over wounds to clean them. It is also somewhat sedative. Rub on skin as insect repellent and perfume, or chew a few leaves to help with illness or to freshen breath.

BURDOCK
Arctium spp.

Family: Asteraceae

Other names: Edible burdock, beggar's buttons, louse bur, *bardane*, wild burdock, wild rhubarb, cuckoo button

Look-alikes: Common cocklebur (*Xanthium strumarium*), creeping thistle, spear thistle, nodding thistle, and other species of burdock, such as woolly burdock (*A. tomentosum*), and teasel (*Dipsacus fullonum*)

Related species: Lesser burdock/common burdock (*A. minus*), greater burdock (*A. lappa*)

WARNING: Pregnant women and diabetics should not use unless working with a skilled herbalist. When harvesting seeds, be careful about inhaling the tiny hairs, which are said to be toxic. Some people have allergic reactions.

> FORAGER NOTE: Not to be confused with the similarly named docks, which are in the *Rumex* genus. Both have large, low-growing leaves, but the stalks, flowers, and seeds are totally different.

Edible and Useful Plants

Description

This nonnative biennial forms a basal rosette in its first year consisting of many large, undulating, elephant ear–like leaves that can reach up to about 2' long and 1½' wide. Leaves are dark green with woolly undersides. Lower leaves are heart shaped or ovate. Smaller leaves also grow densely along the stalks and are alternate and ovate. Margins are often wavy. A deep taproot can reach 1'–2' deep.

In the second year, a stout, erect, grooved, branched stem emerges. The plant flowers from midsummer to fall. Flowers are purple but barely emerge from the rounded, barbed, or hooked bracts that surround them. Flowers with surrounding pointed bracts look like thistles. When the seedpods dry out, they do not change shape.

RECIPE

Gobo

Gobo is a traditional Japanese recipe made with burdock root.

Harvest and clean one large burdock root. The skin should be removed, but it is very thin. Use a potato peeler or knife to gently scrape off the skin. Slice into matchstick-size pieces (about 2 cups worth). Place in a bowl of water, and agitate so that the water becomes cloudy. Change water, and repeat several times until it becomes clear. Add a drop of vinegar and soak while preparing the rest of the dish.

Slice a small carrot into matchstick-size pieces and set aside.

Heat 1 tablespoon sesame oil in a skillet to medium high. While the oil is heating, drain burdock root and then add to hot oil. Stir-fry for 2 to 4 minutes. Add carrots and toss in. Stir-fry for 3 minutes or so. Add 1 tablespoon mirin rice wine vinegar, 2 tablespoons sake, 1 tablespoon sugar or palm sugar, and 1½ tablespoons soy sauce. Stir-fry until liquid cooks off.

Serve hot or cold topped with a sprinkle of sesame seeds.

Arctium lappa grows somewhat taller than *A. minus* and can reach up to 6½' high. Flowers are 1"–1½" in diameter. *A. minus* grows 1'–5' tall. Its flowers are a little smaller, about ½"–1" wide. Also see woolly burdock (*A. tomentosum*), which is similar but with a spider web–like appearance as if covered in wool.

Common cocklebur (*Xanthium strumarium*), another large-leaved, burred herb, has maple leaf–shaped or ovular leaves and more football-shaped or tubular (not round) seedpods, which are similarly covered in hooked bracts.

Range and Habitat

Burdock grows throughout the United States, especially in disturbed areas, forest edges, roadsides, and ditches.

Comments

The leaves, roots, stems, flowers, and seeds are edible. Roots can be eaten raw or cooked and are best in fall of the first year or the following spring (before the plant flowers). Best to harvest when the ground is moist. After it rains, you will have a much easier time digging out the long roots. Use a big shovel. Roasted roots can be used like coffee and can also be dried.

Young leaves and stems are eaten raw or cooked. Stems are best peeled, and older stems are better cooked.

Burdock is widely used for a broad range of medicinal purposes.

CHAMOMILE
Matricaria recutita (Matricaria chamomilla)

Family: Asteraceae
Other names: German chamomile, scented mayweed, wild chamomile
Look-alikes: Pineapple weed (often confused with chamomile because of the similar-looking prominent yellow central disc, but pineapple weed is smaller and does not have the white ray petals)
Related species: Scentless mayweed/scentless chamomile (*M. inodora*, *M. perforata*, *Tripleurospermum perforatum*), stinking chamomile (*Anthemis cotula*), which is **poisonous**
WARNING: Not safe for pregnant women. Can cause contractions of the uterus. A ragweed relative, it can cause an allergic reaction, especially in those allergic to ragweed. It has been reported to cause anaphylactic shock, although most people enjoy chamomile with no problems. Mild sedative effect.

Do not confuse with the very similar-looking stinking chamomile (*Anthemis cotula*), which is **poisonous and has a strong, unpleasant smell**.

Description
This nonnative Eurasian annual looks similar to but much smaller than Shasta daisies with feathery leaves. Lowlying clumps of lovely, daisy-like, white flowers

> ### RECIPE
>
> **Chamomile Mint Tea**
>
> For a pot of tea, harvest several flower heads and a few sprigs of wild mint. Wash and place in tea strainer. Boil water and let cool for a few minutes, and then pour over fresh herbs. Place lid on top of strainer and let steep 10 minutes. Add generous amounts of fresh local honey, and stir well. Serve hot, or enjoy lukewarm throughout the day. A sun tea can also be made with wild chamomile.
> **Variation:** Use dried herbs instead of fresh.

on branched stems. Ray petals are white and surround an upward-pushing, prominent, yellow central disc.

Leaves are bi- to tri-pinnate, or, in other words, they are very deeply lobed into thin, wispy leaf divisions. Leaves appear delicate and feathery. Stems are erect, branched, and smooth, growing 6"–2' tall.

Scented mayweed smells similar to honey or apples. Scentless chamomile is not strongly scented.

Edible and Useful Plants

Range and Habitat

M. recutita is found sporadically throughout the Rockies. Sunny, disturbed areas throughout the temperate zone. *M. inodora* is found from Alaska and across Canada down to Nevada, Utah, and Colorado and across some of the northeastern states. Does not appear to be found in California or Oregon. Look-alike but inedible stinking chamomile (*Anthemis cotula*) is found widely across the United States and has a strong smell.

Comments

Well known as a relaxing tea that is good for digestion and as a mild sleep aid. The stalk, leaves, and flowers can be used. Chamomile has a variety of medicinal uses and can be made into cosmetics such as shampoo and oils. Always steep tea covered to retain the beneficial aromatics.

Scent is an important identifier when making a positive identification of wild chamomile. Make sure to rule out the poisonous look-alike first.

RECIPE

Chamomile Eye Pillow

Harvest 1 cup chamomile flowers and dry in a dark place.

Find a nice soft fabric with relaxing colors, and cut it into two 9" x 4" rectangles. Position fabric so the sides you want on the outside are both facing each other. Place the rectangles so that the edges match up. Sew the two long edges and one short edge using a tight stitch. Pull the sewn edges through so that the pillow is no longer inside out. Gently fill the pillow through the open end with the dried flowers. Sew the remaining edge closed.

At bedtime, lie on your back in a comfortable position that allows your shoulders, neck, and back to relax. Place the eye pillow over your eyes and focus on breathing deeply in and out. Allow the gentle scent of chamomile to waft over you as you drift to sleep.

CHICORY
Cichorium intybus

Family: Asteraceae
Other names: Wild endive, French endive, succory, blue sailors, coffeeweed
Look-alikes: Blue flax, especially from afar

Description
This nonnative biennial or perennial is scraggly, somewhat branched, and grows 1½'–3' tall. Toward its base, there are lots of large, unlobed, dandelion-like leaves up to about 8" long. Leaves become much smaller upward along the stalk.

Known for its large taproot.

Iridescent blue flowers (rarely pink or white) scattered along the thin stalks, often bobbing in the wind. Inflorescences are up to about 1½" across. Flowers mid- to late summer and into fall. Each petal has five tiny grooves along the blunt outer edges, creating a small wave effect.

Range and Habitat
From British Columbia across North America to Florida. Roadsides and disturbed areas.

RECIPE

Garlic Fava Beans and Chicory Greens

Prepare 1 cup fava beans by soaking in water overnight. Rinse well and then simmer (do not boil) in a large pot until soft. Drain and set aside. Retain the bean water.

In a large pot, combine 2 cups water, 5 cloves garlic (chopped), and 2 bay leaves. Bring to a boil. Add 2 cups chopped chicory leaves and reduce heat to a low simmer. Simmer for 10 minutes. Add the cooked fava beans and simmer 5 to 10 more minutes. Remove from heat, and toss with salt and pepper to taste.

Variation: If the leaves are too bitter, first boil in clear water for 5 to 10 minutes, and then drain and proceed with recipe.

Comments

This edible relative of the endive is listed as a noxious weed in Colorado.

To cultivate chicory to mimic store-bought, light-colored endives, cut the plant short and cover with a basket to prevent sunlight from reaching the plant for photosynthesis. Once leaves appear, they will be blanched (white) and tender.

> **RECIPE**
>
> **Chicory Root Coffee**
>
> Strong-tasting dark beverage, usually consumed hot. Harvest the huge roots and cut into small chunks. If the pieces are too large, you will have a hard time grinding them later. Roast in oven at 350°F for 90 minutes to 2 hours. Or roast on a lower temperature for a longer time. Grind the root finely as you would coffee beans. Can be ground in a sturdy blender or food processor. I have read another method of drying the roots by air or on very low temperature, then grinding them, then pan roasting in a dry cast-iron pan. Make your chicory root coffee in a French press and fix as you would coffee.

Roots can also be dug up, dried, and used as a diuretic for medicinal purposes or as a strong tea or coffee-like (but caffeine-free) drink. Green leaves can also be eaten. Many people find them bitter, which can be reduced by boiling the leaves in a few changes of water.

Edible and Useful Plants

CUTLEAF CONEFLOWER
Rudbeckia laciniata

Family: Asteraceae
Other names: Tall coneflower, green-head coneflower

Look-alikes: Black-eyed Susan, sunflower
WARNING: Do not use during pregnancy.

Description

Tall, showy native grows 3'–8' tall. Hairless, erect stems with large leaves that are deeply lobed and coarsely toothed, forming really interesting shapes. Leaves can be large, up to about 12" long, with three to seven graceful, lobed cutouts. Usually found in clusters that form from underground rhizomes.

Large, yellow ray flowers surround a greenish-yellow (sometimes brownish) center button of disc florets. Yellow petal-like ray flowers surround the central disc. Flower heads are up to about 3" in diameter.

Range and Habitat

Mountains of the Four Corners states and Wyoming and into Canada and the eastern United States. Not found wild in the West Coast states. Grows in especially moist areas with rich soil, such as stream banks and ditches, from about 5,000'–8,500' in elevation or higher.

Comments

Especially attractive to birds and bees, including honeybees.

Generally thought to be more actively medicinal than black-eyed Susan. Roots used as a substitute for echinacea. Make a decoction of the roots, and drink as a medicinal tea at onset of illness.

Few sources claim that young shoots, stalks, and leaves are edible raw or cooked. Most people do not have experience eating them, and I have not confirmed their edibility.

Flowers can also be made into a yellow dye.

NOTE: This is not well known as an edible plant. Many consider it medicinal only. I recommend much more research before deciding whether to use it and how to prepare it.

RECIPE

Medicinal Tea

Consult a skilled herbalist.

DANDELION
Taraxacum officinale

Family: Asteraceae
Other names: Common dandelion
Look-alikes: Salsify (flowers), cat's ear (leaves, but cat's ear leaves are hairy), and chicory (leaves)

Description
Dandelions are thought to be introduced, but some accounts say there were pre-Columbian varieties on this continent as well.

This very common perennial grows about 2"–2' high. A sturdy taproot, 3"–8" long, produces a basal rosette of leaves, each 2"–16" long. Taproot can become branched as it ages.

Leaves are oblong and extremely variable. They are quite a bit longer than they are wide. Leaves can be decumbent or erect. They can be lobeless or lobed, often very deeply lobed. Lobes can be rounded or jagged.

Solitary, yellow ray flower clusters of 100 to 300 tiny flowers (that together look like one terminal flower head) stand atop smooth, hollow stems. Flower

> ### RECIPE
>
> **Fresh Greens Salad**
>
> Eating hardy greens raw in a salad is delicious and super nourishing. Like kale, dandelion greens are delicious when massaged with olive oil prior to eating.
>
> Prepare an oil-and-vinegar salad dressing. Combine 3 tablespoons of olive oil and 1 tablespoon of vinegar or the juice of 1 fresh lemon or lime, and sea salt and pepper. Mix well by shaking in a jar until emulsified.
>
> Wash and then chop 2 cups dandelion greens. Place greens in bowl, and pour dressing over them. Mix and massage or toss the dressing into the greens. Set aside and allow to sit for 30 minutes. Just before serving, toss in about 2 cups lettuce of your choice.
>
> **Variation:** Add pan-fried strips of grass-fed steak for a hearty steak salad. Top with chia or other seeds and a few toasted pecans.

heads are 1"–2" wide. Yellow rays are longer toward the outer edges, and most are delicately jagged at the flattish outer tips. Flower color is more or less uniform but can be somewhat darker toward the center of the inflorescence.

Flowers turn to white fluff balls that are assisted by the wind in seed dispersal. Dandelions contain white, milky latex that will ooze out when the plant is cut.

Range and Habitat

Everywhere, especially disturbed areas, burned forests, old fields and pastures, roadsides, gardens, lawns, and avalanche zones from sea level to 13,500' in elevation. Thrive in rich soils and in full sun, but survive in a very wide range of habitats.

Comments

A true hero of the edible plant world! Leaves, flowers, buds, young stalks, and roots are edible raw or cooked. Dandelions are bitter, but often pleasantly so. Best to pick leaves before the flowers appear. Leaves in shady spots will also be more tender and less bitter. Leaves are somewhat less bitter when young but can be eaten when older as well. Tight, unfurled crowns, at ground level in the middle of the rosette, are also tender and edible. To best enjoy the flowers, remove them from the green bracts, although this isn't necessary.

Roots are best eaten when they are young enough to be uniformly whitish throughout, except for the darker skin. Older roots become tougher and darker in the center. Both old and younger roots can be roasted and used for dandelion "coffee."

RECIPE

Bacon Fried Young Greens

This is a hearty breakfast, good before a day of serious lumberjacking or backpacking.

Harvest 1 cup young dandelion leaves. Wash in cool water to remove any dirt if needed. Chop roughly and set aside.

Cook some good-quality bacon in a heavy skillet. When bacon is almost done, add the dandelion greens. Toss until wilted.

Remove from heat.

Serve on a big bed of warm rice with a dash of tamari and a sprinkle of flaxseeds. Top with a fried egg and fresh dandelion flower petals.

Used to make tea, tonics, and salads and in cooked dishes. Boiling and other cooking methods can help reduce the bitterness. Mixing dandelions with other foods, especially strong-tasting foods, helps make them really enjoyable.

Dandelion wine is a well-loved tradition. Harvest fresh flowers and pull the yellow petals from the green bracts to avoid bitterness. Use wine yeast and follow a wine-making recipe.

Dandelions are available from the very beginning of spring to the very end of fall. One of the first flowers to open in spring and, therefore, one of the first foods available each year for both insects and humans. Keep this in mind and, when harvesting in early spring, leave plenty of flowers for the pollinators. Roots can be harvested throughout the winter.

Important food source for cattle, sheep, grouse, gophers, bears, deer, elk, and bees.

While many people frantically poison their dandelions to preserve a pristine grass mono-scape, I just wait patiently and eventually the neighborhood deer swing by and make short work of it in no time at all.

FLEABANE
Erigeron spp.

Family: Asteraceae
Other names: Fleabane daisy, fleawort
Look-alikes: Asters, daisies, *symphyotrichum*
Related species: Many, including featherleaf fleabane (*E. pinnatisectus*), Canada fleabane (*E. canadensis*), common fleabane (*E. philadelphicus*)

Description
This group of common perennial natives has about 170 species in North America. They are difficult to tell apart, but all are low growing and often found in spreading clusters. Fleabanes have short, hairy stems and can grow up to 3' tall for some species. Often seen in the range of 4"–8" high.

Leaves are small, alternate, and lance shaped but vary with species. Some species have leaves that clump toward the bottom of the stalk, while others climb the stalk more evenly. Featherleaf fleabane has more feathery, deeply divided leaves.

Flowers consist of fifty to one hundred narrow lavender or purple petals arranged in a sunburst shape around a yellow center.

> ### RECIPE
> **Doggie Sachet**
>
> Harvest stalk, leaves, and flowers. Use scissors to snip the stalk neatly at the bottom. Gather 10 flowers with stalks and leaves together as if you are holding a bouquet. Secure together using twine. Wind the twine around the base of the stem about four times. Leave about ½" of stem below the twine. Tie the ends of the twine together in a knot to secure. Hang upside down over your dog's bed to repel fleas and other insects.

Range and Habitat

Roadsides, fields, and openings up to about 13,000' in elevation for some species. Common fleabane is found from the Yukon throughout most of the United States. Featherleaf fleabane has a limited range and is found in the mountains of Wyoming, Colorado, and New Mexico.

Comments

While fleabane is generally not considered toxic, I have not found much information on its edibility. I use it for ornamental purposes and hope it keeps the fleas away, not that I've really ever had a flea problem. Some reports say that leaves and sprouts are edible cooked; however, there are so many varieties of fleabane that I urge you to learn more about the particular species you wish to consume before doing so. Fleabane is incredibly common throughout the Rockies, and you will see it frequently.

Can be dried and burned as a smudge to cleanse the house and repel insects. Can be made into an oil for the same purpose.

Many look-alike species. Known for repelling fleas and other bugs.

GOLDENROD
Solidago spp.

Family: Asteraceae

Related species: Rocky Mountain goldenrod (*S. multiradiata*), baby goldenrod (*S. nana*), Canada goldenrod (*S. canadensis*), giant goldenrod (*S. gigantea*), Missouri goldenrod (*S. missouriensis*)

WARNING: Some people are allergic to goldenrod. Try very small amounts at first.

Description
Erect perennial herb from 7" to 2' tall. (*S. gigantea* reaches more than 8' tall.) Stems can be green or reddish and are often branched. Usually smooth, in some species the stems are hairy. Leaves can also be smooth or fuzzy, depending on species. Often grows in small or large, dense patches, giving it a look similar to a yellow-flowered shrub similar to a baby rabbitbrush. On closer inspection, though, it is obviously several flower stalks growing densely together rather than a single shrub.

Leaves are alternate and linear and can be smooth or serrated along the margins.

Tiny yellow flowers with a central disc and outer rays form rounded or pyramidal tufts at the top of the stems. Flowers are often densely clustered, but in

Canadian goldenrod, for example, they can appear leggier and form a less dense cluster.

Range and Habitat
Widely dispersed throughout the continent. While various species are restricted to certain regions, together they cover much of North America. Found along roadsides, disturbed areas, woodland openings, alpine meadows, and tundra. Baby goldenrod is found in a more limited range: from Idaho and Montana to

> RECIPE
>
> **Goldenrod Seed Chicken Soup**
>
> Make a good homemade chicken soup by simmering a whole farm-fresh chicken carcass for 24 hours. Strain and place liquid back in the pot or slow cooker.
>
> Add goldenrod seeds (to thicken soup) along with sliced carrots, celery, garlic, onion, bay leaves, and shredded and diced chicken meat. Season with salt, pepper, and fresh herbs such as thyme, parsley, or rosemary. Simmer for an hour. Serve hot with rice, bread, or a hot potato.

New Mexico and Arizona. Rocky Mountain goldenrod is found in the western half of the United States and throughout Canada.

Comments

There are about seventy-five species of goldenrod. Flowers, seeds, leaves, and fruits are edible raw, cooked, or dried. Also can be used to make dye by soaking and boiling flowers.

PEARLY EVERLASTING
Anaphalis margaritacea

Family: Asteraceae
Other names: Western pearly everlasting

Description
This native perennial sports a tiny, delicate, papery-looking flower that is notable by its showy, rounded, white flower heads. A whorl of papery involucre bracts

> **RECIPE**
>
> **Potherb**
>
> The idea of a potherb is one that has almost been lost to history. Think of it as herbs you put in a pot, hence the name. The tiny leaves of the pearly everlasting are edible but generally are not substantial enough to turn into a meal or even a side dish. They are still packed with nutrients and a tiny dose of flavor, so harvest and add a little to salad, stew, or casserole.

> **RECIPE**
>
> **Simple Sautéed Greens**
>
> Such a delicate-looking plant (although it is quite hardy) requires a simple, delicate recipe.
>
> Briefly sauté young or old greens in a small amount of olive oil over medium heat. Remove from heat, and add a pinch of sea salt. Serve hot.

forms a flower head about ⅜" wide, often with a yellow or brownish center. Leaves are single, slender, and alternate with an obvious midvein along an erect stem that stands 4"–36" high (usually on the shorter side). This lovely plant flowers in early spring to summer, and the delicate white flowers stay almost perfect looking for months.

Range and Habitat
Found widely across the United States from Alaska to New Mexico in mid-high elevations in the mountains. Does not grow in the southernmost states. Usually seen in low-growing clumps along roadsides or disturbed areas and in forest openings. Can grow in poor soil, sand, and loam. Very cold tolerant. Hardy to minus 43°F.

Comments
Young and old leaves are edible, tender, and delicious raw or cooked. The greens have a pleasant, understated flavor. For crafters, the "everlasting" nature of the flower makes this a great addition to dried-flower bouquets and wreaths. Pearly everlasting is also used as incense and as a medicinal astringent. The leaves can be consumed raw or cooked.

PINEAPPLE WEED
Matricaria discoidea

Family: Asteraceae
Other names: Disc mayweed
Look-alikes: Feathery leaves and yellow flower can be confused with chamomile. Pineapple weed is much smaller and lacks the white, daisy-like petals of chamomile.
WARNING: Some people are allergic to this plant. If you are allergic to ragweed or other asters, take caution. Has a mild sedative effect on some people.

RECIPE

Backyard Sun Tea

Fill a clear-glass mason jar ⅓ full with herbs. Tea can be made with pineapple weed alone or add other backyard friends like mint, raspberry leaves, and wild rose flowers. Add water and a lid. Let steep in the sun for 1 to 3 days.

Description
Small, pungent annual ground cover grows 2"–16" high but is more commonly 4"–8" high. Pineapple weed has hairless, branched stems and alternate, feathery leaves ⅛"–2" long. Inflorescences are pineapple scented, yellow or light greenish, cone-shaped discs that do not have petals.

Range and Habitat
This Eurasian native now grows from Alaska to New Mexico on disturbed, compacted soil. Likes full sun. Blooms in spring to late spring and withers in the heat of July. Will sometimes reemerge in August as the weather cools.

Comments
The leaves are sweet and leave the tongue feeling a bit minty. The flower heads are delicately pineapple flavored. Often found as a garden weed.

RABBITBRUSH, COMMON
Ericameria spp.

Family: Asteraceae
Other names: Golden rabbitbrush, chamisa, chamiso
Related species: Rubber rabbitbrush/gray rabbitbrush (*E. nauseosa, E. nauseosus*), Parry's rabbitbrush (*E. parryi*), green rabbitbrush (*E. viscidiflorus*)

All were formerly considered part of the *Chrysothamnus* genus but are now classified as *Ericameria*.
Look-alikes: Big sagebrush (from afar)

Description
This sweet-smelling, native perennial shrub grows rapidly up to 7' tall (sometimes 10') and about 7' wide. Woody lower branches with branched green to gray upper branches. Green rabbitbrush (*E. viscidiflorus*) is somewhat shorter, reaching only about 4' in height.

Wide, rounded crowns flower in late summer to fall and are covered in prolific arrays of small yellow flowers.

Leaves are linear and tiny (about $1/25$"). Foliage is green to grayish and varies within the species. *E. nauseosa* is distinguished by the white, felt-like covering on its branches and leaves that help prevent water loss and make this a highly successful arid-region plant.

Range and Habitat
From Montana and North Dakota throughout the Rockies and Four Corners states to Texas. Found in disturbed areas and open meadows in semi-desert and montane sites up to about 10,800' in elevation.

Very common along moderate-elevation roadsides throughout the region. Does well in high-wind conditions. In fact, I could barely get an adequate photo of rabbitbrush, as it always seemed incredibly windy whenever I came upon the plant, usually along open roadsides from New Mexico to Wyoming.

Comments
Chew raw bark or leaves similar to chewing gum.

Flowers can be used to make sun tea or hot tea.

Rabbitbrush has a variety of medicinal uses, and a medicinal tea of flowers and leaves can be used internally or externally. Flowers can be used to make a yellow dye, and boughs can also be used as a natural building material for outdoor structures, such as ceilings for sweat lodges and floors for cabins.

> ### RECIPE
>
> **Rabbitbrush and Artemisia Soak**
>
> Ritual bathing and aromatic soaks are wonderful ways to decompress from our hectic modern lives. The American understanding of aromatherapy has become confused as fake, toxic chemical scents are increasingly sold as aromatherapy in candles, plug-ins, and other such "air fresheners." Avoid those toxic chemicals and learn to nurture and heal with the wholesome scents of fresh or dried wild plants instead. Instill some calming and nourishing bathing rituals into your life.
>
> Harvest several branches of rabbitbrush with fresh yellow flowers. If you like, also harvest several stalks of one of the big sagebrush or the smaller fringed or white sages. Place several boughs of each into the bathtub, and fill with hot water. Keep the bathroom door closed while the tub is filling to retain the aromatic oils. Turn the lights down low, and sink into your very own desert-scented spa. As you soak, you can use the boughs to gently scrub and exfoliate your skin.

What I love about this plant is that it is so big and prolific; you can harvest several branches from any member of the species without harming the plant or the stand. It also reproduces readily from seed.

SAGE, FRINGED
Artemisia frigida

Family: Asteraceae
Other names: Fringed sagewort, pasture sagewort, prairie sagewort, *romerillo del llano*
Look-alikes: White sage is similar in size, color, and scent, though the leaves are quite different.
WARNING: Do not use during pregnancy.

Description
This spreading shrublet has a pleasant sage fragrance. Its most notable characteristic is its wispy, feathery, deeply divided leaves, which form a mat close to the ground 4"–8" high. Leaves are feathery, downy, silver-green or grayish, and are also found along the stem but less densely.

The flower stalks grow 12"–20" from the mat of leaves and produce a panicle of button-like yellow or silvery-yellow flowers. Not likely to flower in drought years.

> **RECIPE**
>
> **Flower Stalk Tea**
>
> Prepare a hot infused tea using flowers from one or two stalks, fresh or dried. Good at onset of a cold or flu. Can be medicinal and is very strong, so make your tea weak, and do not drink too much without consulting an herbalist. Because it is so strongly flavored, most people will adequately self-monitor and only want to drink small amounts.

Fruits are dry achenes. Similar to white sage (*A. ludoviciana*) but with much more feathery leaves.

Range and Habitat
This native perennial is found in sunny high plains, prairies, dry meadows, hillsides, and overgrazed pastures from the Siberia, British Columbia, and the Yukon to Colorado, Arizona, and New Mexico and to Wisconsin and Minnesota.

Comments
This is such a lovely plant. I have a few in my garden and walk by regularly to run my hands along the leaves for a midday connection to the sage smells of the desert Rockies. Use leaves and flowering stalks as tea or wrap with yarn for a smudge. It is said that it can be used to preserve meat. Can be burned to repel insects.

SAGE, WHITE
Artemisia ludoviciana

Family: Asteraceae

Other names: Wormwood, western mugwort, Louisiana wormwood, cudweed sagewort, gray sagewort, mugwort wormwood, prairie sage, white sagebrush, mariola, alcanfor, estafiate

Related species: Field sagewort/Northern wormwood (*A. campestris*), fringed sage (*A. frigida*), Mexican wormwood (*A. mexicana*)

WARNING: Some people may be allergic. **Do not use during pregnancy.**

Description

This deliciously pungent, erect native perennial generally reaches 2'–3' when mature but is also often seen in young patches just a few inches tall. Can grow to 6'–7', but the weak stems will often flop over if unsupported. Whitish or silvery gray woolly down covers the leaves, giving them their characteristic light-green-gray frosted look.

Leaves are linear, up to 4³⁄₁₀" long. Toward the top of the stalk, many tiny, nodding flower heads appear in late summer. They have yellow central discs and together form a sparse, narrow inflorescence. They turn brown or reddish brown as they dry out in fall.

Fruits are small, dry achenes. Plants usually grow in clumps spreading from underground rhizomes.

Range and Habitat

Widespread from the Northwest Territories to Texas, across the West to California and much of the eastern United States. Dry fields, roadsides, gardens, forests, and forest edges.

> FORAGER NOTE: Not to be confused with the *Salvia* genus of sages. *Salvia* sages have square stems and are related to mint and culinary sage.

> ### RECIPE
>
> **Smudge**
>
> Harvest stalks with healthy, fresh-looking leaves. Tie in a bundle, and hang in a dark area with adequate airflow to dry. Wrap bottom third with twine to secure. Burn as a smudge to purify air, cleanse living spaces, begin a meditation session, or to mark new beginnings.

Comments

Leaves, flowers, and seeds can be used as flavoring or for tea. Widely loved as a smudge. Harvest before or during flowering period. Tea can be used as a bitters for a digestive aid and to help with menstrual cramps, inhaled for lung health, and used for other medicinal purposes. It is antifungal and antimalarial. Can also be used as a deodorant and insect repellent.

SALSIFY
Tragopogon dubius

Family: Asteraceae
Other names: Common goat's beard, yellow salsify, oyster plant, western salsify
Look-alikes: Dandelion, grasses, meadow salsify
Related species: Meadow salsify (*T. pratensis*), oyster plant (*T. porrifolius*)

Description
This biennial (sometimes annual) European native looks similar to a tall, scraggly dandelion with slightly larger flowers, although its grasslike leaves are easily distinguishable. In their first year, leaves are grasslike rosettes that grow to about 14" long. In the second year, a smooth flowering stalk 1'–3' high arises, sending out two to thirty smooth branches, each of which terminates in a composite flower head.

Milky white latex that turns to brown exudes from broken leaves. The leaves are long and deeply folded down the center, creating an obvious, somewhat-rigid V shape, similar to grass but deeper. The leaves along the stem are alternate, linear, and clasping (attached directly to the stalk without a petiole).

Flowers in spring or summer. Showy, yellow, dandelion-like inflorescences are born singly and are ray floret composites, with the outer rays longer. Flowers are dandelion yellow (sometimes pale lemon), about 1½" wide. The inner rays are flecked with darker brown markings and surrounded by markedly longer, thin, pointed bracts (leaflike structures just below and cupping the flower head), giving the flower heads a unique look.

RECIPE

Trifecta Salsify Sampler

Harvest flower buds before they have opened. Also harvest young leaves and stalks.

Remove leaves from stalks and make three piles: buds, leaves, and stalks. Can be steamed or sautéed. Sauté or steam stalks, flower heads, and leaves for just a few minutes. Sprinkle salt on top. Serve hot or at room temperature.

Variation: Drizzle truffle oil or hollandaise sauce over final product.

> ## RECIPE
> **Salsify Leaves Pasta Primavera**
>
> Substitute salsify leaves for the pasta in your favorite recipe.
>
> Harvest 2 cups leaves; wash and set aside. Spiralize one zucchini and add to the salsify leaves.
>
> Prepare the following vegetables by chopping into bite-size pieces: 1 yellow summer squash, 1 cup broccoli, and 1 cup cauliflower. Slice in half 1 cup of cherry tomatoes.
>
> Place chopped veggies in a 9" x 13" baking dish. Toss with enough olive oil to lightly coat the veggies; add ½ teaspoon salt. Roast in oven at 375°F until veggies are soft, about 30 minutes.
>
> When roasted veggies are almost done, bring a pot of water to a boil. Simmer salsify leaves and spiralized zucchini for 5 minutes, then drain water.
>
> To serve, line plates or serving bowl with the "noodles." Top with roasted vegetables. Sprinkle with a handful of pine nuts and plenty of fresh-ground black pepper. Add Parmesan cheese if desired.

The look-alike meadow salsify does not have these extended bracts.

Seeds are attached to a white, fluffy pappus similar to a dandelion and become big, white fluff balls in mid- to late summer.

A related species, *T. porrifolius*, has purple flowers and is also edible.

Range and Habitat
From Alaska to Texas in dry, disturbed sites and roadsides. A common garden weed. Found throughout the plains and foothill ecosystems.

Comments
Flowers, buds, leaves, roots, and young stalks can be eaten raw, steamed, sautéed, or baked. Sweet and tender, similar to an artichoke heart or an oyster. Best when young, as the stalk gets hard after the flowers have gone to seed.

This is a common garden weed and definitely worth harvesting.

SUNFLOWER, COMMON
Helianthus annuus

Family: Asteraceae
WARNING: Some people are allergic to members of the Asteraceae family.

Description

This tall, showy, erect native annual stands 4'–12' high or higher. In 2009, *Guinness World Records* recognized a German *Helianthus annuus* as the tallest ever recorded at 26' 4".

Large, yellow (sometimes red or orange) terminal inflorescences are a composite of hundreds of flowers arranged in a large disc-and-ray floret pattern. The outer ring of petals is yellow. The inner ring is brown or brownish yellow. The flower head is smaller than the cultivated varieties, measuring about 5" across. The tiny flowers on the inner part of the inflorescence create a fractal-like spiral pattern. Blooms late summer to early fall.

Seeds of the wild sunflower are generally smaller than commercial sunflower seeds but otherwise look similar, with a blackish outer shell and succulent seed inside.

The leaves are hairy. Lower leaves are often heart shaped and opposite. Leaves along the stalk are more egg shaped or oblong and alternate. The stout, erect stem is branched many times and hairy. Usually found in large stands.

Range and Habitat

Found in open, often disturbed areas in the plains and foothills. Up to about 9,000' in elevation. This common and widespread bundle of sunshine is actually considered a weed in many states.

Comments

Flowers, seeds, and the seed oil are edible. The oils from the seeds are rich in linoleic acid. Young flower buds can be eaten, often steamed. Seeds can be eaten raw, roasted, dried, or crushed and mixed with other foods.

The leaves can also be crushed and used as a poultice and can also be used as tea for a variety of medicinal purposes. Flowers can be made into medicinal tea, and root decoctions are also used as medicine.

Wild sunflower was cultivated by Native Americans, who selected plants for larger and larger seed size. The USDA reports that such cultivation led to an increase in seed size of about 1,000%.

RECIPE

Sunflower Seed Muesli

Preheat oven to 350°F.

Combine oats, nuts, seeds, and spices: ½ cup shelled whole sunflower seeds, ½ cup other seeds or nuts (slivered almonds, cashews, pecans, pumpkin seeds), 1 cup organic, gluten-free oats (not instant).

Add some or all optional spices: ½ teaspoon ground cinnamon, ¼ teaspoon ground nutmeg, ¼ teaspoon ground cloves, ¼ teaspoon ground cardamom, and ¼ teaspoon ground ginger. A few pinches of salt.

Spread out on a baking sheet. Roast in oven for 10–15 minutes.

Add dried fruit to roasted mixture and toss together: 1–2 cups dried fruit (raisins, currants, dates, figs).

Roast in oven until fruit has plumped. Longer for larger fruits, shorter cook time for smaller fruits.

Remove from oven. Serve atop yogurt or as cereal with homemade almond milk.

THISTLE, BULL / SPEAR THISTLE
Cirsium vulgare

Family: Asteraceae
Other names: Scotch thistle, common thistle, prickly vase, fuller's thistle
Look-alikes: Plumeless thistle (*Carduus acanthoides*), which lacks a prominent midrib on leaves; Scotch thistle (*Onopordum acanthium / O. tauricum*)

Description
This tall, biennial, European immigrant has solitary flowers atop spiny branches. Flowers are rose-purplish, lavender, purple, or sometimes white and are 1"–2" in diameter. Inflorescences are similar (i.e., not divided into ray and disc florets). The entire plant, including the leaves, stems, and branches, is covered in spines and thorns all the way up to the bracts (base of the flower head).

The first-year basal rosette is formed by deeply lobed and very spiny leaves 3"–6" long. In the second year, a many-branched stem 5'–6' tall will emerge, with flowers clustering at the tops. The stem and branches are winged with spines. Leaves are more prominent along the stem and its base, becoming progressively smaller toward the top of the stalk.

> FORAGER NOTE: Stems are winged with spines, and the entire plant is covered in them. Distinguishable from creeping thistle because bull thistle is not only covered in spikes but also forms a basal rosette, which its perennial relative does not.

Seeds have a feathery pappus and are wind dispersed. Seeds remain viable for only a few years and usually germinate quickly.

Range and Habitat

Hardy to USDA Zone 2. An early-successional species, the bull thistle requires sun. It does not grow in shade.

Found on disturbed sites including pastureland, overgrazed ranges, forest clear-cuts, roadsides, and ditches from Alaska to Texas. It is uncommon in natural, ungrazed grasslands.

Comments

Flowers, buds, leaves, roots, stem, seeds, and oil are edible but somewhat bland flavored. Good to mix into more flavorful dishes. Young stems can be stripped of their thorns and cooked like asparagus. Leaves can be soaked in salt water and

---

RECIPE

Taproot and Fresh Greens Stew

Carefully harvest starchy taproot. Look for basal rosettes in the fall, and harvest carefully with thick gloves and a big shovel. Remove leaves, and scrub root with water and a stiff brush to remove dirt. Chop root into 2" pieces.

Add to slow cooker with 4 carrots, 1 large yam, and 1 large turnip, all chopped into about 2" pieces. Fill slow cooker ¾ full with water. Add 3 bay leaves, ½ jalapeño, and ½ teaspoon each salt and pepper. Add ½ cup barley. Add brisket, ham hock, or other meat if desired. Cook until bubbling and all ingredients are soft.

To serve, place 1 cup chopped kale in soup bowl. Spoon hot stew on top of the fresh greens. Enjoy as is, or garnish with fresh-chopped herbs such as basil, parsley, rosemary, or mint. Add salt and pepper to taste.

> **RECIPE**
>
> **Lightly Boiled Flower Buds**
>
> Harvest 1 cup young flower buds before they open. Bring medium-size pot of lightly salted water to a boil and reduce heat. Add buds, simmer for 5–10 minutes, then drain. Consider serving as a unique side dish with grilled black bean burgers or New Mexico–style green chili soup.

then cooked. Root can be dried or cooked. Flowers and buds can be cooked. Seeds can be roasted and eaten.

Bull thistle flowers are full of nectar and provide food for honeybees, bumblebees, hummingbirds, and butterflies. Seeds are eaten by American goldfinches, juncos, mice, and many other birds and rodents.

THISTLE, CREEPING
Cirsium arvense

Family: Asteraceae
Other names: Canadian thistle, California thistle, field thistle, lettuce from hell thistle, cursed thistle, prickly thistle, small-flowered thistle, way thistle
Look-alikes: Sow thistle, burdock, Bigelow's tansy aster (*Machaeranthera bigelovii*)
Related species: Bull thistle (*C. vulgare*), Hooker's thistle (*C. hookerianum*), leafy thistle (*C. foliosum*)

Description

This nonnative, creeping perennial, lavender- or pink-purple-flowered member of the aster family reproduces both by spreading root systems and by seed. Inflorescences are nicely scented, up to about 1¼" wide and long. Bracts are not spiny or prickly as so many other thistles are.

Flowers appear in clusters at the top of the stalks. Florets are similar (not divided into ray and disc florets). Creeping thistle flowers in late spring or early summer and sometimes again throughout the season. By late summer or fall, it gives way to a white and feathery pappus (like a dandelion when it goes to seed).

Its erect, branching, smooth stems grow 1'–4' tall. Stems may become hairy as they age. Leaves are alternate and spiny and grow 3"–8" long along the stalk. Leaves are larger toward the bottom and smaller toward the top of the stem.

The extensive root systems will produce new shoots when tilled or broken. They extend 6'–15' deep and can grow more than 15' horizontally. Seeds, especially when deeply buried, remain viable for more than 20 years. Seven hundred to 1,500 (or more) seeds are produced per stem.

Two main factors distinguish *C. arvense* from other thistles: Male and female flowers are on separate plants, and dense patches are formed by the spreading root systems, hence the name creeping thistle. Compared to the spiny-stemmed

bull thistle (*Cirsium vulgare*) and nodding thistle (*Carduus nutans*), Canadian thistle stems are not spiny.

Range and Habitat

Hardy to USDA Zone 4. The common name Canadian thistle is misleading as this species is actually a European native. It is found widely from Alaska to New Mexico and California, as well as to the East Coast where there is moderate moisture. It is considered a noxious weed in most states because it thrives on sunny pastureland and other stressed and disturbed environments, where it decreases range habitat for cattle. A 1998 study showed that in Colorado alone, about 400,000 acres were "infested" with Canadian thistle.

Does not survive in shade or healthy, diverse ecosystems.

Hummingbirds use thistle fluff to build their tiny nests. Honeybees also rely on creeping thistle and are a main pollinator of it.

Comments

Roots, leaves, stems, and young flowers are edible. Young, still-flexible stalks can be eaten raw or cooked. Especially good to harvest when the plant is 1' tall or less. The roots can also be used as food and medicine. Despine leaves before eating. Roots can be used medicinally to support the liver.

This is a very common weed throughout the United States. The plant reproduces readily, so feel free to harvest as much as you want. Many farmers, private citizens, and government entities regularly poison stands of thistle, so harvesting for your own use will also help prevent the continued use of poisonous herbicides. Of course, make sure you are gathering *prior* to any chemical eradication efforts.

THISTLE, NODDING
Carduus nutans

Family: Asteraceae
Other names: Musk thistle, nodding plumeless thistle, *chardon penché*, plumeless thistle, thistle, milk thistle

Description
This biennial thistle produces a rosette of narrow, spiny leaves 11"–23" long in spring or summer of its first year. By mid-spring of the second year, branched stems will emerge from the rosette. In warmer climates the entire life cycle will sometimes be completed in 1 year. Leaves are waxy, dark green, and coarsely lobed, with sharp yellowish, brownish, or whitish spines at the tips of the lobes.

Reaches 3'–5' tall. Stems are hairy, spiny, branched, and smoother toward the top. The top portion of the stems and branches may host a few spines that are much smaller than those along the bottom two-thirds of the stalks. The very spiny leaves also become sparser, or absent, toward the top.

> FORAGER NOTE: Flower heads are notable by the very large bracts below the flowering portion that resemble an artichoke. The name "nodding thistle" comes from another notable characteristic: Instead of standing erect, the flower heads typically nod downward up to 90 degrees or more, especially as they mature.

Spherical flower heads 1"–3" in diameter are notable by the very large bracts below the flowering portion.

Somewhat resembles its relative the artichoke.

Each plant produces anywhere from one to fifty flower heads in late spring to late summer. Each flower can produce over a thousand seeds, which are fluffy, dispersed by wind (although they often don't make it very far), and can remain viable for more than a decade. Flower heads droop 90 to 120 degrees once mature, and they are less droopy when young.

Range and Habitat

From British Columbia to Texas and from coast to coast. Found from sea level to about 8,500' in elevation or higher. Nodding thistle grows on disturbed sites, meadows, and pastures with moderate moisture. Prefers full sun. Does not grow well in shade.

Comments

Can be eaten like other thistles. Peel stem and eat the young, still-soft part, but not the woody inner stem. Considered helpful for liver function.

RECIPE

Boiled Stem

Remove spines as described for other thistles. Gently peel the outer layer of the young stem. Simmer for a few minutes until soft. Sprinkle with truffle oil or butter and a bit of tamari or sea salt.

YARROW
Achillea millefolium

Family: Asteraceae
Other names: Soldier's woundwort, knight's milfoil, milfoil, common yarrow, bloodwort, carpenter's weed, *hierba de las cortaduras, plumajillo*, gordaldo, nosebleed plant
Look-alikes: Poison hemlock (*Conium maculatum*), water hemlock (*Cicuta maculata*), osha (*Ligusticum porteri*), osha del campo (Angelica; *Angelica grayi*), Queen Anne's lace (*Daucus carota*), wild parsley (*Lomatium* spp.), water parsnip (*Sium suave*)
WARNING: Do not confuse with the deadly poison hemlock or water hemlock. Yarrow leaves are distinct and with practice are very recognizable and distinguishable from the more parsley-looking leaves of the poisonous species.

Description

This perennial member of the aster family is similar to some members of the carrot family in that its tiny flowers and florets form white, umbrella-shaped clusters. Wild varieties are mostly white, but pink and yellow varieties are cultivated. Some cultivated varieties are huge, reaching 5' tall.

Conspicuous white umbels top an erect stem, 8"–2' tall. Fernlike leaves are lacy and compact. They are alternate along the stem. Basal leaves and those

Edible and Useful Plants   109

lower on the stem are larger than those higher up. Leaves are 2"–8" long. Flower heads are 2"–5" in diameter.

Native, with some introduced species. Yarrow flowers from spring well into late summer. After flowering, the fruits are achenes and retain much the same look as the original flower.

Range and Habitat
Yarrow grows from northern Canada to Texas and is found on sunny slopes, along roadsides, and in forest openings and disturbed areas up to 11,400' in elevation or so.

Comments
The Latin name is derived from Achilles, the Greek warrior who used yarrow poultices to stop his soldiers from bleeding to death. It was, and still is, also

> **RECIPE**
>
> **Poultice to Stop Bleeding**
>
> Yarrow can be part of your wild plant first-aid kit. The leaves of yarrow can be chewed or mashed by hand to make a poultice. Place the mashed leaves on an open wound or a large gash, and pack the bleeding opening with yarrow leaves for much faster clotting time.

widely used throughout the Rocky Mountain region as a bitter tea and can be used as a poultice to stop bleeding.

All parts of the plant can be used for tea, but it is best to harvest the leaves and flowers, not the root, so that the perennial root system remains intact and can provide next year's harvest.

Dried yarrow makes a sturdy ornamental flower—a great reminder of summer that can last for years.

Rub yarrow leaves on the skin for a natural bug repellent.

Harvest by removing one leaf from several different plants. Leaves can be dried and stored for winter use as tea or for your first-aid kit to stop bleeding.

> **RECIPE**
>
> **Yarrow Tea**
>
> Yarrow leaves and flowers can be made into a tea and taken internally or poured over a wound or open sores. Tea for both external and internal use can be made from fresh or dried leaves and flowers. To dry, harvest fresh leaves (and flowers if you like) and hang to dry in a dark place. When crisp, store in a jar through winter.
>
> For a delightful, mellow-tasting tea, use a few leaves per mug of tea. Bring water to a boil, and let it cool for a couple of minutes. Pour over yarrow leaves. Cover and steep for 5 minutes. For a digestive tonic, drink ¼ cup of the warm or cooled tea slowly before mealtime. Tea can be stored in the refrigerator for several days. Excellent mixed with ginger root, mint, and honey.
>
> Yarrow leaf tea is one of my favorite wild teas. It is easy to find yarrow, fairly easy to identify, and the taste is delightful. Pick a couple of leaves per plant, and spread your harvesting out. Leaves stay fresh throughout the growing season so you can harvest any time of year. In no time at all, you can harvest enough to last through winter. And because yarrow often grows in large stands, it is easy to spread the harvest out and not harm the stand.

BERBERIDACEAE / BARBERRY FAMILY

OREGON GRAPE
Mahonia repens

Family: Berberidaceae
Other names: Creeping Oregon grape, barberry, creeping mahonia, creeping barberry
Look-alikes: Holly
Related species: Tall Oregon grape / tall mahonia / holly-leaved barberry / Oregon holly grape (*M. aquifolium*)
WARNING: Roots are medicine and should be avoided during pregnancy, while breast-feeding, and if you have an overactive thyroid. They contain high doses of berberine, which can cause vomiting, kidney infection, and other medical problems.

Description
Native perennial shrubs or subshrubs. The wild mahonias can be tall (*M. aquifolium*) or low-lying (*M. repens*). Creeping mahonia (*M. repens*) forms low-lying, spreading clusters 4"–12" tall. Tall mahonia grows 1'–10' in height.

FORAGER NOTE: Though sometimes called barberry, this plant should not be confused with the many species of shrubs also called barberry, which are in the *Berberis* genus. Also not bearberry/kinnikinnick (uva-ursi).

Leaves are green or blue-green, stiff, leathery, and waxy. This shrub is considered an evergreen, but the leaves turn reddish as they age, and in the fall and winter, they create a sturdy, colorful ground cover. Leaves are pinnately compound, with five to eleven opposite, oval-shaped leaflets per leaf stem and one single leaflet at the tip of each leaf stem. Leaflets have spiny-looking teeth spaced out along the leaf edges.

Dense clusters of sweet-smelling yellow flowers give way to juicy bluish or purple berries about the size of blueberries or somewhat smaller.

Edible and Useful Plants 113

> ### RECIPE
>
> **Mahonia and Watermelon Juice**
>
> Collect 1 cup berries and place in blender. Add water just to cover berries. Blend well. Strain through a colander, collecting the juice in a bowl below. Place juice back into blender with 6 ice cubes and 2 cups chopped watermelon. Add honey if desired. Blend until combined.
>
> Serve in chilled glass with a pinch of cinnamon and chopped fresh mint.

Range and Habitat

Creeping Oregon grape is found from British Columbia to Colorado and Texas and throughout the West to California; it is found spottily in the northeastern United States. Tall Oregon grape is more a plant of the Northwest and is found from British Columbia and Alberta to Idaho, Oregon, and California, as well as sparsely throughout the northeastern United States. Grows in forests, forest edges, dappled shade, and sometimes full sun. Found in low to mid-elevations.

Comments

These little berries range from sour to delicious. They can be eaten raw, cooked, or dried. I have heard that in the wet climate of the Northwest, they are much sweeter than in the dry climate of the Four Corners region. I have found delicious berries from the creeping mahonia in Colorado, even in summers of drought, though oftentimes I will find many shrubs with no berries at all. If you find a sweet, juicy one, you'll be very happy, though it's more common for them to be quite sour. This is also a common garden ornamental.

The bright-yellow flowers are also edible. Can be made into jam or jelly.

Berries can be made into a beverage like grape juice, fermented into wine, or made into jam. The stem, root, and leaves are used for medicinal purposes.

RECIPE

Oregon Grape Flower Popsicles

In a saucepan heat 6 cups water just to a simmer. Add 1½ cups bright-yellow fresh Oregon grape flowers. Stir and simmer for about 10–20 minutes. Remove from heat and add ¾ cup honey or sugar. Stir until dissolved. Place in refrigerator until fully cold.

Pour into Popsicle molds and freeze.

Variation: Add uncooked flowers into Popsicle molds along with liquid tea. Can use Oregon grape flowers and add whatever other edible flowers you like: lilac, dandelion, or for a weird and spicy addition, try nasturtium.

BORAGINACEAE / BORAGE FAMILY / FORGET-ME-NOT FAMILY

BLUEBELL, MOUNTAIN
Mertensia ciliata, *Mertensia* spp.

Family: Boraginaceae
Other names: Mountain bluebells, tall fringed bluebells, streamside bluebells
WARNING: Reports of stomachache if too much is eaten

Description
The mountain bluebell is a native perennial. It is an herbaceous flowering plant with long stalks that grow 1'–4' tall. I generally see it on the taller side in my area, though there are several species in the Rocky Mountains, including a shorter variety.

Clumps of leafy stems produce clusters of bell-shaped, blue, tubular flowers. Flower buds emerge pink and become a striking blue color as they mature.

Flowers are ½"–1" long. Tubular corolla surrounds five petals per flower with a flaring out toward the ends.

Leaves along the stalks are oval or lance shaped with a pointed tip. They are alternative, with lateral veins. Leaves are generally larger toward the bottom and smaller toward the top of the stalk.

Rhizomes spread underground, forming extensive patches of mountain bluebells.

Range and Habitat

Mountain bluebell is found in moist soils near streams and wet meadows, usually in partial shade. Found between 5,000' and 12,000' altitude. Associated with Engelmann spruce, subalpine fir, and aspen.

Comments

The flowers and leaves are edible raw or cooked. Harvest flowers from early to late summer when in bloom. When leaves are older or hairy, they are less palatable and may be improved with cooking. When you find a large stand of bluebells, harvest a few leaves and a few flowers from multiple plants.

While many flowers are sweet in flavor, the mountain bluebell is more umami flavored. A surprisingly hearty and delicious trail nibble.

RECIPE

Bluebell and Chickpea Fresh Summer Salad

Flowers are always a crowd-pleaser to top any meal. A fresh summer salad is an excellent way to celebrate this beautiful edible.

Use ½ cup mountain bluebell flowers; set aside for the very end.

Pan-fry 1 can chickpeas, drained and rinsed, in olive oil with a few pinches of salt until partially crispy. Set aside.

Combine salad ingredients: fresh lettuce, ½ red onion (chopped), 1 pint cherry tomatoes (halved), 1 yellow pepper (chopped), 1 cucumber (sliced and halved), ½ cup fresh parsley (chopped). Add to taste: fresh mint, sweet cicely, dandelion leaves.

Dressing: Combine 2 tablespoons red wine vinegar, 1 tablespoon olive oil, and juice of 1 lemon. Mix well.

Toss all ingredients together (except the flowers), then top with bluebell flowers.

BRASSICACEAE / MUSTARD FAMILY

BITTERCRESS, HEART-LEAVED
Cardamine cordifolia

Family: Brassicaceae
Other names: Heartleaf bittercress, toothwort
Look-alikes: Watercress, chickweed, hairy bittercress (*C. hirsuta*), and other bittercress species

> **RECIPE**
>
> **Bittercress Salad**
>
> Wash if needed and dry in a salad spinner. Combine bittercress with a big helping of garden-fresh lettuce. Toss with olive oil, vinegar, and sea salt, and top with dried cranberries and toasted pecans.

WARNING: Often grows in or near stagnant water where parasites, such as liver flukes and others, can live. Be mindful of water sources and the potential need for cleaning bittercress before eating it.

Description

When this plant emerges, it first sends out tiny, purple stems shortly after the snow melts, which soon turn green. Upright stalks extend up to 30" tall. Often grows in dense patches of hundreds of individual stalks spreading by underground runners.

Showy tufts of white flowers emerge at the top of the stalks and continue to flower through summer. Flowers have four white petals, each about ¼"–¾" long.

Leaves are alternate, heart shaped, and undivided (similar) with shallow, wavy lobes. Prominent veins radiate from the stalk to the leaf margins. White balls or calluses are sometimes seen where the leaf veins meet the leaf margin. Leaves are up to about 4" long.

Range and Habitat

This native, herbaceous perennial can be found from British Columbia to Northern California, east through Colorado and Wyoming, and south to New Mexico. Typically found in or near streams, moist areas along lakes, alpine meadows, wetlands, or moist forests, especially in shade or partial shade.

There are about 170 species of *Cardamine* (bittercress) throughout the world and about fifty in North America. Range is from 2,000' to 12,000' in altitude.

Comments

The flowers and young leaves of this mustard family species are edible and taste similar to mustard or horseradish with a strong mustard or peppery flavor. Eat raw or cooked. Can be added to salads, soups, stews, casseroles, and omelets, or use them as you would any flavorful green. This tasty wild mustard green is a perfect addition to a salad or can be gently cooked.

RECIPE

Bittercress Easy Omelet

Chop ½ cup bittercress into smaller pieces. Heat olive oil in a fry pan on low to medium heat. Add the greens and toss until wilted. Add 2 farm-fresh eggs, break the yolks, and scramble together. Sprinkle sea salt and pepper on top. Continue to stir until the eggs are done. Remove from pan and enjoy.

PENNYCRESS
Thlaspi arvense

Family: Brassicaceae
Other names: Frenchweed, stinkweed, fanweed, field pennycress
Look-alikes: Shepherd's purse

Description
This introduced annual member of the mustard family grows 6"–2½' in height. It is hairless with lanceolate, alternate leaves and slightly wavy margins that can have blunted teeth.

Small, four-petaled flowers are white and form open or wispy, rounded clusters toward the top of the stalk.

Obvious rounded seedpods are light green and flat and often have one small, narrow indentation on the outer edge. Seedpods are thin discs with a large lump in the middle where the seeds reside inside.

Range and Habitat

From Alaska to Florida along roadsides, disturbed areas, open fields, and in gardens. Plains to montane zone.

Comments

Seeds and leaves are edible raw or cooked. This is a common garden weed and one to harvest.

The leaves are similar in flavor to cultivated mustard greens and are tender, spicy, and delicious. The seeds are a mustard substitute and very similar to, although often far more flavorful and spicy than, grocery store mustard. Eat the entire seedpod before it dries out, or crack open and harvest the seeds after the pod dries. Seeds can be used to flavor any dish, crushed to make mustard, or soaked in olive oil to make a flavored oil. Use flavored oil for dressings or cooking.

RECIPE

Pennycress Coconut Soup with Bok Choy and Tofu

In a large saucepan or soup pot, combine one 13.5-fluid-ounce can coconut milk, 3 cups water, and 2 vegan bouillon cubes. Add a dash of fish oil and a dash of tamari sauce. Add 1 teaspoon palm sugar or brown sugar. Also add 1 chopped Thai pepper and 2 tablespoons pennycress seedpods (fresh, not dried). Turn heat to medium.

Add 2 medium-size purple potatoes and 1 medium-size carrot, both sliced into ¾" cubes. Bring mixture to a simmer. Add 2 cups cubed raw chicken. Simmer for 15 minutes, stirring occasionally, with lid slightly ajar, until chicken is cooked through and potatoes are tender.

Meanwhile, chop 2 cups bok choy greens and stalks; set aside. Once potatoes and chicken are cooked, add bok choy to pot. Simmer covered, stirring occasionally, for 6 to 8 minutes. Add salt and pepper to taste. Serve hot.

SHEPHERD'S PURSE
Capsella bursa-pastoris

Family: Brassicaceae
Look-alikes: Pennycress
WARNING: Can induce uterine contractions, so do not ingest during pregnancy. Avoid if you are taking medication for high blood pressure or have thyroid or heart problems. Possible sedative.

Description

This annual (sometimes biennial) Eurasian immigrant is similar to other mustard family members. It has a cluster of basal mustard or dandelion-like leaves that are deeply lobed. Smaller, unlobed, pointed leaves are along the stalk.

Flowers are small and white, with four petals, forming a very loose raceme along the upper portion of the stalk. They appear to be dancing up the stem toward the sky. Seeds are encapsulated in light-green seedpods similar to those of pennycress but triangular or heart shaped rather than rounded. Mature plants are sometimes quite small but can reach 12"–18" high. Flowers and seedpods often appear on one plant at the same time.

Edible and Useful Plants

Range and Habitat
Disturbed areas and fields throughout the plains and foothills of the Rocky Mountains. Widespread throughout the continent. Common garden weed.

Comments
This wild mustard family edible is well loved in China and Korea. All parts of the plant are edible. Use greens cooked or in salads. There is great variation in the taste of the seeds. Sometimes has a strong mustard flavor, and other times it has no flavor at all. When flavorful, seeds and seedpods make a great mustardy flavoring, and the roots can be used like ginger.

RECIPE

Wild Greens Wontons

The mustard flavor is a great complement to soy dipping sauce, green onions, and shrimp. To make wonton filling, combine 2 cups diced shrimp or tofu, ½ cup diced green onions, ½ cup diced baby bok choy, ½ cup diced mushrooms, 1 egg (beaten), ½ teaspoon fresh-grated ginger, 1 tablespoon tamari, ¾ tablespoon sesame oil, and ½ teaspoon organic cornstarch with ¾ cup diced shepherd's purse leaves and/or green seedpods (use less if using the spicy seedpods rather than greens). Mix together well.

Have a bowl of flour and a cup of cool water ready. Use pre-made wonton wrappers (or make your own). Place a small spoonful of mixture into center of each wonton wrapper. Wet the edges lightly with water. Fold over so that edges meet; squeeze and tuck the sides gently so they stick together. Dip and roll gently in flour, and place on cookie sheets. Can be cooked immediately or frozen (place cookie sheet into freezer, and once wontons are frozen, store in sealed containers).

To cook, bring a large pot of water to a rolling boil; add wontons. Stir very gently so that the wontons do not stick together. When wontons float to the top, they are ready to serve.

Variation: Fry wontons instead of boiling.

CACTACEAE / CACTUS FAMILY
Forage for cacti with thick gloves!

PRICKLY PEAR
Opuntia spp.

Family: Cactaceae
Related species: Plains prickly pear cactus (*O. polyacantha*), brittle prickly pear cactus (*O. fragilis*)
Other names: Nopal, nopales, paddle cactus, tuna cactus, Indian fig
WARNING: Large thorns are obvious. Tiny thorns are not so obvious. Gloves are recommended.

Description

This native evergreen perennial consists of about 200 species of prickly pear cacti. Different species readily hybridize.

Clusters of flat, rounded, fleshy stems (cladodes) that are like paddles or beaver tails make up the body of the prickly pear. Leaves are tiny and short-lived. Many species are low shrubs, found in expanding clumps, although some have woody trunks with the green paddles clustering toward the top. Paddles are mucilaginous inside, similar to aloe.

Tiny, barbed, thornlike hairs (glochids) often go unnoticed, as the larger spines are more obvious. Avoid both. The small hairs can become lodged in the skin for several days.

Large, showy flowers are usually yellow but sometimes pink, reddish, or purple. Below the flowers, in late summer or fall, emerge fleshy fruits whose size and sweetness vary with species. Fruits are pink or red, sometimes green, yellow, or purple, with seeds inside.

Range and Habitat

Deserts and dry grasslands from British Columbia to Texas and Mexico.

Comments

All members of this species are edible, though the bigger, fleshier varieties are more enticing. For example, the brittle prickly pear cactus (*O. fragilis*), as the name suggests, is not one of the better edibles. The sweet, red fruit, often called tuna or pear, is the most desirable part of the prickly pear. The mucilaginous vegetable portion, called the nopal, or pad, is found inside the green paddle-like stem. It also is edible and delicious, tasting somewhat like tart green beans.

Flower petals and seeds are also edible. When harvesting flower petals, take only a few. Leave the reproductive parts intact to allow the fruit to form.

RECIPE

Grilled Nopal Salsa

Remove prickly skin from the green paddles. Lightly coat inner flesh with oil. Grill gently until just browned. Remove from heat and dice. Combine with equal parts diced, raw, garden-fresh tomatoes. Add chopped onion, chopped garlic, cilantro, and plenty of fresh-squeezed lime juice. Salt and pepper to taste. Serve with grilled chicken and organic corn chips.

> **RECIPE**
>
> **Nopal Fruit Salad**
>
> Prepare fruit by stripping the skin. Cut into quarters or, if very large fruits, bite-size pieces. Prepare 2 cups of the cactus fruit. Slice other fruit such as peaches, watermelon, cantaloupe, blueberries, or plums. Cut all into bite-size pieces. Combine all fruit and toss gently. If you like, squeeze fresh lemon or lime juice over the fruit and toss together gently. Top with pine nuts or garden-fresh mint. Serve immediately.

Nopales should be harvested when they are firm, not wrinkled. In some species, the fruits are big, sweet, and fleshy; in others, they are less so. Fruits are made into juice, jam, and wine or eaten raw. Pads also can be eaten raw, grilled, or sautéed.

To harvest the fruit or pad, use a sharp knife and make a clean slice about 1" above where it attaches to the pad below. Carefully peel the outer layer off the fleshy interior by making one cut down the length and then removing strips of skin with the knife. It is advisable to wear thick protective gloves. To harvest the pads, you must remove the small spines, or glochids. The skin can be stripped or left on. Spines can also be scrubbed off with a stiff brush, burned off, or rolled in sand to remove.

Prickly pear is a great survival food because of its high water content and, if needed, can help stave off dehydration.

The mucilaginous inside portion of the paddle can also be used as a lotion or salve.

> **RECIPE**
>
> **Prickly Pear Juice**
>
> Harvest several fruits. Slice them in half. Using a spoon, remove the seeds, and then scoop out the fleshy centers. Place 2 cups fruit in blender with 1 cup water. Blend on medium speed until pulverized. Enjoy as is, or combine with sparkling water to make a spritzer.

CAMPANULACEAE / BELLFLOWER FAMILY

HAREBELL
Campanula rotundifolia

Family: Campanulaceae
Other names: Bluebells of Scotland, bluebell bellflower
Related species: Parry's bellflower (*C. parryi*), rampion bellflower (*C. rapunculoides*)

Description

This lovely native perennial in the bellflower family produces lavender-colored, bell-shaped flowers. Flower heads are five-lobed and dangle downward like a bell from delicate, erect green stalks. Stalks grow to about 4"–40" tall, usually on the smaller side when I have seen them. Often two or more flowers droop in loose clusters from each stalk.

Leaves are denser toward the bottom of the stalk, becoming sparser toward the top. Leaves are like small, delicate blades of grass

> **RECIPE**
>
> **Harebell Leaf and Flower Salad**
>
> Cut a few leaves from the stalk and add to a salad with a few purple flowers. The leaves are small, and an all-harebell leaf salad is probably impractical, especially considering its vulnerability. But the flowers are sweet, and I think edible flowers always make a meal better.

attached to the stem to about halfway up and can remain green even after early frosts. Basal leaves are rounded and serrated and can be useful in identifying the plant before it flowers. The rounded leaves usually show themselves early and wither before you find the flowers blooming.

Range and Habitat
Edges of woods or grassy areas, dappled shade, or full sun. Hardy to USDA Zone 3 and found from Alaska to Texas and throughout the United States except for the Southeast.

Comments
Harebells grow in clusters but usually in small ones. They are considered endangered or vulnerable in some states. It is important to consider this plant a delicacy and harvest only very small amounts—but a delicacy it surely is.

Leaves, raw or cooked, and flowers are edible. The flowers are like fresh forest candy, and just one will turn your day around. There's just no way not to be happy after sampling a sweet harebell flower. The root can be chewed for medicinal purposes (heart and lungs).

The roots are also edible and are best enjoyed cooked. When harvesting, be sure to leave plenty of roots in the ground to allow for regrowth.

> **RECIPE**
>
> **Romantic Vanilla Cake with Harebell Blossom**
>
> I am a sucker for a good vanilla cupcake with a hint of almond extract and buttercream frosting. It's a perfect addition to a romantic mid- to late-summer picnic with your special someone. Pick the vanilla cupcake recipe of your choice. Bake and let cool, and frost with homemade vanilla icing. Top with just a few purple harebell flowers, and enjoy. For extra romance, add a few rose petals as well.

Edible and Useful Plants

CAPRIFOLIACEAE / HONEYSUCKLE FAMILY

ELDERBERRY
Sambucus nigra

Family: Caprifoliaceae
Other names: Blue elderberry, black elderberry (black and blue elderberry are two different species)
Look-alikes: Mountain ash (the entire shrub or tree is generally larger than the elderberry, with somewhat larger leaves and berries; berries are orange or orange-red), red elderberry (*S. racemosa*)
WARNING: Seeds of the fruits are toxic when raw and need to be neutralized by drying or cooking. Do not eat berries with seeds raw, though seeds can be

removed in a Foley mill, and the pulp can be eaten raw. Avoid the red elderberry, which has either red or orange berries. Some sources say the red elderberry species is toxic; however, there is conflicting information. Stick with the black, blue, and purple berries. The stems, bark, root, and leaves are also toxic and should not be eaten.

Description
This native perennial shrub or small tree is often seen in clusters and can grow up to about 16'–20' tall but is often seen shorter, around 4'–5'. The elderberry has pinnate leaves with five to nine paired, opposite, serrated leaflets and one lone leaflet at the tip. Leaflets are long and thin and are lanceolate or ovate in shape. They are hairless, smooth, and glossy.

Aromatic, small white or cream-colored flowers, with five petals and five stamens each, form in dense, branched, flat-topped or pyramid-shaped clusters, or cymes, spreading up to about 5.9" across.

Edible berries are found in the blue and black elderberry varieties. The berries are round, blue or black, and sometimes with a powdery coating that makes the darker berries appear bluer.

Range and Habitat
Variations of *S. nigra* are found throughout the continental United States and some parts of Canada. Prefer moist, well-drained, and sunny habitat. Often found in seral communities (intermediate stages of ecological succession) and in moist forest openings, along streams, on slopes, and in canyons. Often associated with serviceberry, chokecherry, gooseberries, big sagebrush, and some grasses.

Comments
Flowers and berries are edible, but berries need to be either cooked or dried. If eating raw, remove the seeds before consuming. Do not eat the stems or leaves.

Flowers can be eaten raw or cooked. Harvest flowers in spring and berries in late summer or fall. Both can be harvested by snipping off the entire cluster with scissors or a knife. Pick the blooms or berries from the stem at home and process.

RECIPE

Elderberry Cold and Flu Syrup

Cover 1 cup fresh or dried elderberries with 4 to 5 cups water. Add several slices fresh ginger, 1 to 2 cinnamon sticks, and ½ teaspoon whole cloves. Bring to a boil and lower to a simmer. Cover and simmer gently for about 45 minutes until liquid has reduced and mixture thickens.

Remove from heat and mash the berries up a bit with the back of a large wooden spoon so that when you strain the liquid out, you'll get all the juicy goodness. Strain through a fine strainer, pressing the berries with a wooden spoon.

While the liquid is still warm but not super-hot, add raw honey to taste. Don't go overboard, but make it yummy.

At the first onset of cold or flu, adults should take up to 1 tablespoon about every 3 hours.

Up the ante with this recipe by adding potent osha root. Add a piece of osha root about the size of your thumb to the mixture, and simmer osha right along with the rest of the ingredients.

Berries are excellent dried or for use as pie, jelly, or wine. Flowers can be fried into fritters or used to make liquor. Save dried berries for tea or a tonic for winter colds and flu. Fresh or cooked berries can be deseeded in a Foley mill.

Bears enjoy eating the berries, and moose, elk, and deer browse on the leaves and stems. Berries are also eaten by the western bluebird, grouse, pheasant, common house finch, red-shafted flicker, ash-throated flycatcher, Steller's jay, western tanager, and many other birds.

Medicinal uses are beyond the scope of this book; however, I will mention that elderberry is particularly high in vitamin C and is commonly used as an antimicrobial, antiviral, immune support and to help speed recovery from cold and flu. The recipe above is an excellent concoction to take at the onset of cold or flu symptoms.

CHENOPODIACEAE / GOOSEFOOT FAMILY
Recognizable by leaves shaped like the foot of a duck or goose.

LAMB'S-QUARTER
Chenopodium album

Family: Chenopodiaceae
Other names: Wild spinach, goosefoot, white goosefoot, wild quinoa
Look-alikes: Strawberry blite, hairy nightshade (*Solanum physalifolium*), ground cherry nightshade (*Solanum villosum*)
WARNING: Contains oxalates. Seeds and seedpods should be rinsed to remove saponin.

Description

This annual herb has toothed, rounded-triangular, or goosefoot-shaped leaves that are arranged alternately along gently ridged, erect, often-branched stems. From above, they appear whorled. Leaves are soft and fleshy and, especially when young, have a whitish crystalline residue emanating from the leaf base. Leaves hang from slightly drooping petioles and are ½"–3" long.

This fast-growing annual ranges from a few inches to 4'–5' tall (sometimes up to 6½'), depending on soil and moisture conditions.

Flowers are small, pale green, and round with no petals.

As the days shorten or water becomes scarce, lamb's-quarter begins to set seed. First, tight buds form along the tops of branches. In early fall, you will begin to see seeds popping out of the seed capsules. Seeds are flat, mostly black but also brown or reddish, and about 1 millimeter across, or about one-fourth the size of commercial quinoa seeds.

Range and Habitat

A prolific pioneer species that colonizes disturbed areas and grows well in all soil types all over the world. In the Rocky Mountains, it grows up to 11,000' in elevation or higher.

Comments

Leaves, stalks, and flowers are edible raw or cooked. Seeds are also edible sprouted or cooked. It's a delicious and prolific wild edible pioneer.

This very common garden weed is an excellent source of food. Lamb's-quarter, along with dandelions, should help us all transform our negative weeding mindset into a happily abundant harvesting mindset. For some reason, our conquer-and-control culture despises plants that voluntarily join our garden

RECIPE

Rice Bowl with Wild Greens

One of my favorite breakfasts is a rice bowl with an egg and lots of fresh veggies. Prepare sticky rice or your rice of choice in advance.

I have a patch of lamb's-quarter in my yard so it's easy to go outside and harvest a handful of leaves. I also have arugula, parsley, mint, chives, and sweet cicely growing in spring and they all go so well together. Wander in the garden and harvest greens and herbs. Chop roughly and place in a bowl with rice.

Fry or scramble an egg with salt, garlic powder, and paprika. Add to bowl. Top with a sprinkle of flaxseeds, a drizzle of olive oil, and tamari.

> ### RECIPE
>
> **Wild Spinach Salad**
>
> Harvest leaves before the plant flowers. Large and small leaves are good as long as they look fresh. Mix with other early lettuce greens and baby chard.
>
> To make dressing, in a separate bowl combine 1 tablespoon olive oil, 1 teaspoon balsamic vinegar, 1 heaping teaspoon tahini, 1 clove fresh garlic (chopped), sea salt, and fresh-ground pepper to taste. Mix well with a hand blender or in a bullet blender, until well combined. Pour over salad. Toss until all greens are covered. Top with crisp tofu squares, meat, or fish, and 1 teaspoon flaxseeds.

ecosystems. These volunteers are not our enemies, and if we can relax a bit and learn to appreciate and be nurtured and nourished by them, life becomes a bit easier. Wild edibles are our caretakers, and wild spinach is a prolific one.

Its leaves provide tender, delicious greens from early spring on. They can be eaten raw or cooked. They can also be dried and turned into a green powder and added to smoothies throughout the winter. Their production tapers off in the heat of summer, but some young plants will reappear by fall. Because lamb's-quarter reseeds so readily, you may want to tame the stands in your garden. Harvest or compost seeds to prevent an all-out takeover, but I highly recommend leaving a good stand for food.

Seeds are time-consuming to winnow after harvest. The green capsule should be removed. It contains saponin and is bitter. The capsules can be toasted and then the seeds crushed out and winnowed, or you can just let them dry out and then crush and winnow the seeds. Winnowing works best when there is a smooth, steady breeze.

STRAWBERRY BLITE
Chenopodium capitatum

Family: Chenopodiaceae
Other names: Beetroot, strawberry spinach, Indian-paint, Indian ink, blite goosefoot
Look-alikes: Lamb's-quarter
WARNING: Leaves contain oxalates, as does spinach. Seeds may be toxic in large amounts. Also contains saponin, which can be toxic.

Description
This annual member of the goosefoot family is erect, sometimes decumbent (lying down but with branches reaching upward), growing to about 2' tall. In poor conditions, it is much smaller, about 6" high. Varies in size dramatically, much like its relative lamb's-quarter. Size is dependent on growing conditions.

Strawberry blite is notable by its bulbous light-green flowers that cluster along the stalk and become weird-looking, bulbous, bright-red fleshy fruits. They somewhat resemble strawberries but grow clustered along the stalk and are about ¾" wide (sometimes much tinier).

Leaves are alternate, from very small to about 3" long. They are broad and triangular, with wavy margins or deeply lobed, bluntly pointed teeth. Leaves resemble those of lamb's-quarter.

Range and Habitat
From Alaska south through New Mexico. Also found in the northern states and eastward across Canada. Prefers moist soil and sun, though I have seen it growing in partial shade and in dry soil.

Comments
Leaves and fruits are edible raw or cooked. Leaves can be used similarly to lamb's-quarter.

The first time strawberry blite appeared in my yard, I had absolutely no idea what it was. As the name suggests, it looks like a bulbously bright-red disease taking over an otherwise vibrant plant.

I've read accounts of this species that say the fruits are bland and the seeds problematic. I don't find this to be the case. I think these fruits are sweet, soft, and delicious. There are probably variations within the species, but I have found that the seeds are easily chewed, much like raspberry seeds, just adding a little texture. I would actually rate this up there with thimbleberry as one of the most delicious yet least-known wild fruits, with the added awesomeness that the plant also provides tender greens. The only problem is that it's not very common and does not reseed readily, so harvest thoughtfully.

RECIPE

Blite Salad

Harvest 1 cup strawberry blite greens. Wash and spin in a salad spinner. Combine in a large salad bowl with 3 cups fluffy, lighter salad greens, such as red curly lettuce. Add 1 cup watercress.

Chop all greens to bite-size pieces. Toss well with 1 tablespoon olive oil, ½ tablespoon fresh-squeezed lemon juice, ½ teaspoon apple cider vinegar, and salt and pepper to taste. Add the red fruits of strawberry blite and 2 tablespoons crushed cashews. Enjoy this fresh, crunchy, sweet, spicy, and fruity salad.

Variation: Substitute slivered almonds for cashews.

CRASSULACEAE / STONECROP FAMILY

ROSEROOT
Sedum integrifolium and *S. rhodanthum*

Family: Crassulaceae
Other names: King's crown (*S. integrifolium*, *Rhodiola integrifolia*), queen's crown (*S. rhodanthum*, *R. rhodantha*), New Mexico stonecrop, Leedy's roseroot, ledge stonecrop, *Rhodiola neomexicana*, *Sedum roseum*, *Sedum rosea*

Description
Native succulent perennial with broad, flat, thick leaves that crowd the stem. Leaves are fleshy and ovate, can be smooth or toothed, and vary in size by subspecies, typically less than 1" long. Erect stems 1"–9" tall grow in clusters. Leaves are alternate along the stalk but appear whorled, as the alternate rows are not evenly lined up. Leaves are green, sometimes with a bluish-white powdery coating.

Inflorescences are terminal (at the top of the stalk) clusters of pink, rose, or dark-red flowers. They are flattish to rounded. Flowers bloom in mid- to late summer. Red flowers are king's crown; pink flowers are queen's crown.

> **RECIPE**
>
> **Warm Roseroot with Spinach**
>
> Harvest 1 cup of young shoots and leaves. Chop. Heat 1 tablespoon mild-flavored olive oil in skillet over medium heat. Toss in the roseroot shoots and leaves and cook gently for a few minutes. Add 1 cup spinach and toss to wilt over heat. Then remove pan from heat. Add sea salt and pepper to taste.

Range and Habitat

Roseroot prefers cool, high-altitude locations. It is found from Alaska and Newfoundland to the mountains of New Mexico and in some locations in the eastern United States. Grows in moist areas along rocky slopes, open meadows, creek banks, and high-altitude riparian zones. Found in the arctic and in alpine and subalpine zones from about 9,500'–12,000' in elevation.

Comments

Leaves, shoots, flowers, and roots are edible raw, pickled, or cooked. The juicy leaves can provide water in survival situations and are high in vitamins A and C. Slightly bitter tasting; use similar to greens or asparagus. Similar to stonecrop but taller and not quite as delicious to eat but still very good.

Several varieties are considered endangered or threatened because of warming temperatures or habitat destruction. Keep this in mind, and harvest only when you find a large, healthy stand.

Used medicinally for stress relief.

STONECROP
Sedum lanceolatum

Family: Crassulaceae
Other names: Lance-leaved stonecrop, spear-leaf stonecrop, yellow stonecrop
Look-alikes: Wormleaf stonecrop (*S. stenopetalum*) is also edible.
WARNING: Some sources warn of possible stomach upset when large quantities are consumed.

Description
This native perennial is short and succulent, with showy yellow flowers. Hairless, erect, succulent stalks about 2"–8" high. Fat, linear, alternate leaves grow closely along the stalk, creating a whorled appearance. Leaves are small, up to about 1" long, and also form a basal rosette. Stalks and leaves are either green or pink-red.

Bright-yellow buds become bright-yellow flowers at the top of each stalk. Flowers have five narrow, yellow petals, usually seen in short clusters.

Range and Habitat
Found from the Yukon to New Mexico. Open, rocky hillsides from sea level to the subalpine zone.

Comments

Called lance-leaf stonecrop to distinguish from its relative, roseroot, which is also a succulent but has flattened leaves. Leaves and shoots can be eaten raw, pickled, or cooked. Roots can be roasted or boiled. The plant is mucilaginous, much like mallow or okra, and has a pleasant cucumber-like flavor.

Leaves can be used as a poultice for bites and rashes by mashing them into a ball. Tea can be made from the leaves, stems, and flowers. This tea is said to be used to clean out the womb after childbirth. Some sources also report that stonecrop has some laxative effect.

> ## RECIPE
>
> **Salty Fish Salad**
>
> Harvest young stalks and leaves of stonecrop. Chop into small bite-size pieces.
>
> Combine in a bowl with smoked oysters (drained if from a can) or smoked trout cut into bite-size pieces. Add finely chopped parsley and a touch of olive oil and vinegar. Toss together gently.
>
> **Variations:** Add cherry tomatoes cut in half, and toss together with rest of mixture. Add chopped red onion.

CUPRESSACEAE / CYPRESS FAMILY

JUNIPER, COMMON
Juniperus communis

Family: Cupressaceae
Other names: Common juniper, dwarf juniper, prostrate juniper, mountain juniper, old field common juniper, ground juniper

Description
Native spreading evergreen shrub or small tree. In the Rockies, common juniper is most often seen as an undulating, mat-forming ground cover or shrub that can reach about 4'9" in height (often shorter) and 13' wide. Needlelike leaves are stubby and green.

The small cones (less than ½") look like round seeds or berries and can be easily recognized as juniper "berries." Young

FORAGER NOTE: Leaves are like needles, not cedar-like (i.e., not scaled). Obvious berries, usually year-round.

Edible and Useful Plants   143

berries are green and tender, turning dark blue, purplish, or blue-black with maturity at around 18 months.

Range and Habitat

The USDA Forest Service says this might be the most widely distributed tree in the world. From Alaska and northern Canada to New Mexico and around the globe.

> **RECIPE**
>
> **Holiday Necklaces**
>
> Pick a bowl full of juniper berries. Using a strong needle and thread, carefully press needle through berries and make a necklace as long as you like. It's easiest to make it long enough to fit over your head and just tie the ends of the thread together. Use berries of varying color, from light green to dark purple, for an interesting pattern. Add some rose hips for variation.

Comments

Food for deer, mountain goats, and other species, especially in winter and early spring. A popular home landscaping staple.

It grows vigorously and often gets out of control, taking up too much room in less than expansive gardens. Most people have seen this or related varieties in home gardens across the United States but ignored its edible and medicinal qualities. As long as you do not use poisons on your lawn or garden, common juniper can make a great start to your year-round backyard pantry.

These berries are strongly flavored and can be eaten raw but are more like a breath mint than a blueberry.

RECIPE

Lemonade Cocktail with Fresh Juniper

Soak a handful of juniper berries in a bottle of gin or vodka for several weeks. The alcohol will take on the strong juniper flavor, depending on how many you use. Juniper berries are traditionally used to make gin, and an extra "fresh-hopped" addition of the berries makes an exciting and earthy treat. Mix and serve on ice with sparkling lemonade or other light mixers that allow you to taste the juniper.

Variation: Make a juniper mojito by combining fresh-crushed mint, a bit of sugar or another sweetener, and juniper-infused vodka with fresh-squeezed lime juice and ice. Shake well.

JUNIPER, ROCKY MOUNTAIN
Juniperus scopulorum

Family: Cupressaceae
Other names: Rocky Mountain cedar, mountain red cedar
WARNING: Some accounts warn that this plant can cause kidney failure, convulsions, and irritated digestive tract with overdoses. Pregnant women, children under age 12, people with cancer, and people with kidney disease should **not** ingest.

Description
This native perennial evergreen is an erect shrub that often appears more like a small tree. It forms a dense conical or pyramidal shape that grows 3'–33' tall and when mature is usually about 20'–30' tall. Leaves begin as needlelike, and when mature, they are covered in small scales, earning it the name cedar-like juniper.

Rocky Mountain juniper also has a prolific array of small, hard, powdery blue or purplish ovulate cones about ¼"–⅜" in diameter

> FORAGER NOTE: Cedar-like scaled leaves or needles. Obvious berries, usually year-round.

that most people refer to as berries. Seeds mature in their second year. Green in the first year, they turn deep blue or purple.

Trees live for 250 to 300 years, some for 1,000 years. They begin producing seeds at 10 to 20 years but are most productive from about 50 to 200 years.

Rocky Mountain juniper reproduces by seed and is pollinated mostly by wind. Hybridizes with other species, which may cause some difficulty in identification.

Range and Habitat

Rocky Mountain juniper is found on dry, rocky slopes from British Columbia and Alberta to Texas. Can be used for screens or hedges. Hardy to USDA Zone 3. Requires only 10" of rain per year. Found near sea level in the Northwest and up to 9,000' in the Southwest. Climax species in juniper and piñon-juniper habitats. Common species that is important for both humans and wildlife throughout the region.

Comments

Light-green, blue, or purple berries are edible. Berries are strong-tasting, sort of like peppercorns but with their own strong, unique flavor. Younger, green berries are more astringent. Older, blue berries are sweeter. I nibble on both while on

> ### RECIPE
>
> **Raw Juniper Berries**
>
> One of the most rewarding ways to eat wild foods when possible is raw and straight from the plant. It allows you to experience the plant and understand its properties in a pure and powerful form that just doesn't happen when it gets mixed up in a fancy recipe.
>
> Hiking along any trail in our region, you will pass one form of juniper or another. I prefer to pick the green tender berries and the older blue ones. See warning above, and don't over-consume.
>
> **Variation:** Eat 1 juniper berry together with 1 rose hip. (Seeds are edible.) Both can be found hanging from bushes year-round.

the trail. This is not the kind of berry, such as blueberries, that you would eat a big bowl of. It's more of a flavoring or spice to be used in small quantities, about one to three berries at a time. It can also be chewed like a mint to freshen breath.

Juniper is well known as the flavoring ingredient in gin. The berries are extremely flavorful, and a small amount goes a long way. Can be harvested year-round at any time during the berry's life cycle. Can be used as a spice similar to pepper or rosemary, cooked with meats or fish, or used as medicinal tea. Use them fresh or dried.

Rocky Mountain juniper provides dense, protective shelter for wildlife and migratory birds, including chipping sparrows, robins, song sparrows, mockingbirds, sharp-shinned hawks, juncos, and myrtle warblers; also mice, voles, and wood rats. Large game animals also utilize Rocky Mountain juniper for forage and protection.

> ### RECIPE
>
> **Simple Lamb or Venison Stew with Crushed Juniper Berries**
>
> Combine in a slow cooker: 2 pounds lamb or venison with bones, 12 crushed fresh or dried juniper berries, 4 carrots (chopped), 4 stalks celery (chopped), 3 potatoes (cubed), ½ cup black-eyed peas, ½ teaspoon turmeric, 5 bay leaves, and 1 jalapeño (or less to taste). Fill slow cooker with water.
>
> Cook on medium or high for several hours. Once beans are soft and fully cooked, add salt and pepper to taste. Add additional water during cooking if needed.

ELAEAGNACEAE / OLEASTER FAMILY

CANADA BUFFALOBERRY
Shepherdia canadensis

Family: Elaeagnaceae
Other names: Soapberry, buffalo-berry, russet buffaloberry, russet red buffaloberry, Canadian buffalo-berry, *Hippophae canadensis*, *Elaeagnus canadensis*, *Lepargyrea canadensis*
WARNING: Fruits contain saponin, which can be harmful if eaten in large quantities.

Description
Native perennial, nitrogen-fixing shrub grows 3'–13' tall. Oval leaves are dark green and opposite, ¾"–2⅓" long. The distinguishing characteristics are the small but prominent rust-colored spots on the under surface of the leaves, along with silvery hairs. Flowers are yellow to brownish. Fruits are red or yellowish berrylike achenes. Fruits ripen from midsummer to early fall. May be stunted (as in the photo) in drought years.

Range and Habitat
From Alaska and northern Canada across the western United States to New Mexico and California and across to the Northeast. Grows in sun or dappled shade.

Comments
Fruit has high saponin content and can, therefore, be used as soap. Fruit can also be eaten raw, dried, or cooked. Berries are sweeter after the first frost. There are also a variety of medicinal uses.

RECIPE

Indian Ice Cream

In a clean glass bowl (not greasy and not plastic), combine 2 cups buffaloberries and 2 cups water. Beat mixture well by hand or with a metal mixer until it becomes foamy, like a soft meringue. Add honey or another sweetener, and beat them in. Serve as a bittersweet dessert.

NOTE: Contact with grease or plastic will prevent froth from forming.

RUSSIAN OLIVE
Elaeagnus angustifolia

Family: Elaeagnaceae
Other names: Russian silverberry, oleaster

Description
This Eurasian native is an invasive deciduous shrub or tree that grows 12'–45' tall. It is a somewhat gangly, bushy tree with distinct powdery, silver-green oblong or linear leaves that are long and thin and sometimes curl into a partial tube shape. Its fruits grow in drupes and are similar to small gray-green olives, although not related to commercial varieties of olives. They ripen in late summer or early fall.

RECIPE

Russian Olive Jelly

Wash fruits and remove stems. Simmer 4 cups Russian olive fruits in enough water to cover them. Simmer for about 30 minutes, mashing the fruit with a wooden spoon. Remove from heat and run through a Foley mill, retaining the liquid and pulp in a bowl below.

Put juice-pulp mixture back into a saucepan. Combine with 1 packet of pectin. Bring to a boil, and then add ½ cup sugar or another sweetener if desired. Stir well. Bring to a rolling boil and constantly stir for 1 minute while boiling. Remove from heat and place in jars; store in refrigerator.

For long-term storage, use canning methods appropriate for jelly. Place in sterilized glass jars with sealed lids, as you would other jellies.

Variation: For a savory jelly, instead of sugar, use fresh or dried hot peppers and a little bit of salt.

The roots are nitrogen fixers, allowing Russian olive to thrive even in very poor soil.

Range and Habitat

Russian olive is a massively invasive nonnative and is now very common along ditches, creeks, and rivers throughout the region from about 4,500'–6,000' in elevation.

Comments

The fruits can be eaten raw, dried, cooked, or made into a beverage. Fruits are astringent and are best when fully ripened. They are also mealy. Can be used as a seasoning or to make sorbet or jelly. High in vitamins A, C, and E and also in essential fatty acids and flavonoids. They can be blended with water to make a beverage that is mildly sweet. Best to strain out or eat around the seeds.

The sale of Russian olive trees is banned in Colorado, and they are considered an extreme nuisance species, so harvest fruits freely. They like riverbanks and riparian areas, use a lot of water, and are accused of crowding out native species including cottonwood (although river diversions and dams have played a significant role in the cottonwood's decline as well). Russian olives now provide shelter and food for birds including pheasants and others.

Not related to the culinary species of olives. The fruits look similar but are smaller than most commercial olives.

EPHEDRACEAE / MORMON TEA FAMILY

MORMON TEA
Ephedra spp.

Family: Ephedraceae
Other names: Joint fir, American ephedra
Look-alikes: Horsetail, scouring rush
Related species: Species include *E. viridis* (most common), *E. torreyana*, and *E. cutleri*
WARNING: If buying Mormon tea in a store, be careful that you are not purchasing Chinese ephedra (also called *ma huang*), which is a *much* more potent drug.

Description
This native evergreen is a small to medium-size green, broom-like shrub growing to 6' tall. Jointed stems are numerous, branching, and crisscrossing. Stems have tiny, scaled leaves that are ⅕" long and barely noticeable. Yellow pollen sacs are prominent in spring. The female plants have green seedpods.

One or several spore-producing cones emerge from the nodes. Leaves and bracts are grouped in two to three, depending on species.

> ### RECIPE
>
> **Mormon Tea**
>
> Harvest plenty of Mormon tea and dry in cool, dark place with adequate airflow. Store dried stems in an airtight jar. The herb will remain more potent if left whole until ready to use, but it also can be ground or chopped into smaller pieces.
>
> Place a spoonful of dried herb into a tea strainer. Add boiling water and steep until desired strength.
>
> **Variation:** Add lemon balm, clover, alfalfa, peppermint leaves, rose hips, or honey.

Range and Habitat

From southwestern Wyoming and western Colorado to Arizona and New Mexico. Also in parts of Oregon and California. Found in desert, dry foothills, mountain grasslands, piñon-juniper and ponderosa pine forests, and river basins.

Comments

Related to medicinal ephedrine varieties but with a much milder effect. Can be useful as a decongestant, opening the lungs, and for allergy symptom relief.

Tea is made from the branches. Used for medicinal purposes but can also be used as a beverage on a regular basis. Tea is high in calcium.

> ### RECIPE
>
> **Trail Chew**
>
> While on the trail, chew on a fresh sprig of Mormon tea. It will feel somewhat astringent in the mouth and has a fairly strong, but not unpleasant, flavor and aftertaste.

EQUISETACEAE FAMILY / HORSETAIL FAMILY

HORSETAIL
Equisetum arvense

Family: Equisetaceae
Other names: Field horsetail, common horsetail
Look-alikes: Scouring rush, Mormon tea, young shoots resemble asparagus
WARNING: Sources advise that large quantities can be toxic because of the enzyme thiaminase, which robs the body of certain B vitamins. Small quantities are fine. Cooking or drying removes thiaminase. Do not gather from heavily fertilized fields or ditches, as this species can concentrate nitrates and become toxic. Also can concentrate selenium when heavy in soils.

Description

Horsetail has two very distinct growth phases. In spring it sends out a fertile, unbranched shoot that resembles a fleshy asparagus shoot. The shoot also somewhat resembles scouring rush, but it is light brown to brown and more fleshy and stout (thicker). It is 4"–6" tall and about ¼" wide. A spore-producing cone (which also resembles the scouring rush) tops the shoot before it withers away. Most people do not recognize this asparagus-like shoot as horsetail because it looks so different than the mature plant.

In late spring, the perennial root systems send out infertile vegetative shoots. These green, jointed stems, ridged at the joints, grow to about 2' tall (sometimes taller). Smaller, similar-looking branches emerge from the main stalks, creating the appearance of an emerging evergreen tree. Small, narrow leaves are whorled at stem joints.

FORAGER NOTE: At certain stages of development, horsetail can look similar to scouring rush (*E. hyemale*), but horsetail is distinguished by the whorled leaves protruding from the stem joints.

Range and Habitat
Found from 2,900' in altitude in Montana up to 10,800' in Colorado. In other regions, it is found down to sea level. Prefers moist sites in woods, fields, meadows, riverbanks, lakes, disturbed areas, and swamps. Also found in drier locations, such as roadsides. From Alaska to Texas and through much of the eastern United States.

Comments
Can be taken for short periods of time as a medicinal tea to strengthen connective tissue. Also used as a diuretic and for liver, kidney, and lung health.

> **RECIPE**
>
> **Horsetail Tea**
>
> Pour boiling water over a heaping teaspoon of horsetail. Steep covered for a few minutes and enjoy. Horsetail can be dried and used to make tea throughout the winter.

SCOURING RUSH
Equisetum hyemale

Family: Equisetaceae
Other names: Common scouring rush, scouring rush horsetail, horsetail, rough horsetail, scouring horsetail, tall scouring rush, western scouring rush, tall scouring rush, Dutch rush, *Hippochaete hyemalis*
Look-alikes: Bamboo, horsetail, Mormon tea
WARNING: Sources advise that large quantities can be toxic because the plant contains the enzyme thiaminase, which robs the body of vitamin B. Small quantities are fine. Cooking or drying removes thiaminase. Do not gather from heavily fertilized fields or ditches (where there might be fertilizer runoff), as this species can concentrate nitrates and become toxic. Also can concentrate selenium when selenium is heavy in soils.

> FORAGER NOTE: At certain stages of development, scouring rush can look like horsetail (*E. arvense*), but unlike horsetail, scouring rush has no leaves.

Edible and Useful Plants 159

> **RECIPE**
>
> **Homemade Toothpaste**
>
> Use dried scouring rush. Blend in blender to a powder.
>
> Combine equal parts powdered scouring rush and powdered clay with a few drops of essential mint or cinnamon oil. Add just enough water to make a very thick paste. Make enough to last for a week's worth of toothbrushing. Store in refrigerator for up to a week.

Description

This perennial native looks like a smaller version of bamboo. Erect, straight, grasslike or bamboo-like stalks stand up to about 3' tall. Hollow, jointed, and leafless. Periodic black-and-white bands encircle the stalk. Spreads by rhizomatous root systems and is usually found in large patches.

Spore cones, instead of flower heads, are dark yellow with black spots. Cones are about ¾" long.

Range and Habitat

Very moist locations from Alaska across the entire Rocky Mountain region to Florida. Moist forests, forest edges, and riparian areas. Can tolerate full submersion in water. Found up to about 9,840' in elevation.

Comments

Stems are covered in silica, which is good for teeth and bones. Blend a stalk in blender with your smoothie, or steep for tea. Often confused with its relative *E. arvense*, also called horsetail. Many medicinal uses for liver, kidneys, and urinary tract ailments, as well as asthma.

Use dried bundles to scour and shine aluminum and copper and to polish wood. Provides food for geese and other waterfowl. Also good for scrubbing pots, pans, and teeth.

ERICACEAE / HEATH FAMILY

KINNIKINNICK
Arctostaphylos uva-ursi

Family: Ericaceae
Other names: Bearberry, Indian tobacco, hog cranberry, *uva-ursi*, beargrape, creeping manzanita, *coralillo*, kinnik-kinnik, k'nickk'neck
Look-alikes: Mountain huckleberry, cranberry, lingonberry, mountain cranberry, creeping snowberry
WARNING: Some sources warn that *uva-ursi* should not be used by pregnant women. It is reported to cause decreased blood flow to the fetus.

> FORAGER NOTE: In late summer and fall, look for bright-red berries in the forest ground cover.

Description
This perennial subshrub blankets forest floors throughout the region. Dark-green leaves are small, rounded, spoon- or heart-shaped with a distinct cleft or vein running their length. Leaves are sturdy,

Edible and Useful Plants

evergreen, and have a leathery or plastic feeling. The shrub reaches a height of 8"–10", though often less.

Small pink or cream flowers give way to very mealy, bright-red berries. The berries can be confused with cranberries and lingonberries but are much mealier.

> ### RECIPE
>
> **Pemmican**
>
> Think about what it would take to really survive off the land without the grocery store or the freezer. You would use what was available and figure out ways to make it last through winter. Pemmican is just that—a mixture of rendered fat (suet), dried meat, dried fruits or berries, nuts, and seeds.
>
> Blend well in a blender or food processor 2 cups dried meat; place in a bowl. In blender, chop 1½ cups dried *uva-ursi* berries and ¼ cup raw pine nuts or sunflower seeds. Add to bowl. Mix in 1 cup warm (just liquid) suet. Massage together well with hands or a wooden spoon. Spread out as a thin layer onto a cookie sheet and allow to cool. Slice into strips and store in a sealed container. Great trail snack throughout the year.
>
> **Variation:** Add honey to mixture.

Bearberries also have several larger seeds inside, which cranberries and lingonberries lack.

Range and Habitat
From Alaska and the Northwest Territories south across western North America through New Mexico. From 3,000' to 9,000' in elevation but beginning higher in the southern reaches of the range.

Comments
Dried leaves are a traditional smoking herb. People commonly refer to *uva-ursi* as kinnikinnick, although the Native American word *kinnikinnick* is said to refer to a smoking mixture usually containing a variety of bark and leaves that often included *uva-ursi*.

The plant is also called bearberry because bears are thought to be fond of eating the berries. Leaves can be used to tan hides. Like cranberries, they are good for urinary tract health.

MOUNTAIN HUCKLEBERRY / BLUEBERRY
Vaccinium myrtillus

Family: Ericaceae
Other names: Grouseberry, whortleberry, bilberry
Look-alikes: Kinnikinnick, cranberry, lingonberry, mountain cranberry, creeping snowberry

Description
White bell-shaped flowers with a tinge of pink become deep-purple berries (red for some species) in mid- to late summer. Light-green leaves grow alternately along a stiff stem and are elliptical or ovate to lanceolate in shape. Berries are about the size of a medium-size blueberry. The berries have a rounded, somewhat starlike concentric circle indent on the bottom. Grows 4"–18" tall.

Species in the *Vaccinium* genus can crossbreed and hybridize. In my area I most commonly see *V. myrtillus* that have deep-purple berries hidden under small leaves. Also explore *V. scoparium*

FORAGER NOTE: In late summer and fall, berries will ripen. They are often hidden under low leaves, so crouch down and look beneath the green ground cover.

(berries are more red, hence the name red whortleberry), *V. membranaceum*, and *V. caespitosum* for related species that can also be found in the Rocky Mountain region.

People often confuse *V. myrtillus* with kinnikinnick as they both blanket forest floors in similar ways. However, the leaves of the wild blueberries are lighter green in color and thinner.

Range and Habitat

From British Columbia and Alberta east to central Oregon and down through Wyoming and Colorado to northern New Mexico and Arizona. Abundant in the southern Rockies. Also found in Europe

RECIPE

Trail Snack

Use in any way that you would use blueberries. Can be frozen, made into jam, smoothies, or pies, or to add a sweet punch to salads. Can also be dehydrated in your food dehydrator or in the sun. Honestly, very few of these ever make it home with me as they are such a perfect trail snack.

and Asia. An understory variety is found in coniferous forests across the Rockies. Often associated with lodgepole pine, ponderosa pine, Douglas fir, and other related species.

Comments

A truly delicious edible berry. You will often find expansive stands of the wild blueberry and so can feel comfortable harvesting. Harvesting does not harm the plant, but be mindful as you walk through and always leave plenty for the wild animals and other hikers who might want to share the experience.

There does not appear to be any real agreement on the common name of this delicious and prolific berry or the various species.

The deep-blue/black color is due to high levels of an antioxidant flavonoid called anthocyanins.

FABACEAE / PEA FAMILY

Seeds look similar to pea pods, but beware: There are many poisonous, inedible species in this family.

ALFALFA
Medicago sativa

Family: Fabaceae
Other names: Lucerne
Look-alikes: Yellow and white sweet clover (before they bloom)
WARNING: Can be problematic for people with lupus and gout. Contains saponin, which in large quantities can cause problems. Avoid during pregnancy and while breastfeeding. The sprouts manufacture a substance called canavanine to protect them from predation. This substance is toxic to humans and can sometimes cause lupus- or arthritis-like symptoms. Monitoring for joint stiffness is recommended.

Description
This long-living, nonnative perennial is erect and gangly. The plant's height depends on the growing conditions but usually matures at about 2'–3' tall. Stalk is branched and generally smooth.

Alfalfa leaves are formed in clusters of three small, oblong leaflets similar to those of clovers, but the leaflets are somewhat more narrow and elongated (ovular). Leaves are alternate along the stalk and not spaced uniformly. The upper half of the leaflets are sharply toothed. There is a noticeable darker indent with a slight fold at the vein, down the middle of each leaf.

The pealike flowers form ovular, rounded clusters of purple or bluish, rose, sometimes yellowish or whitish flowers that are up to about ⅜" long. Forms seedpods much like other members of the legume family but curled like a ram's horn. Seeds are small and kidney shaped. The plant has a very robust root system.

Alfalfa is an important food crop for both humans and animals. It is used as feed for poultry and livestock, and in the wild, it is grazed upon by deer, elk, antelopes, geese, grouse, sandhill cranes, mallard ducks, partridges, pheasants, and others. The leaves are also used commercially to fortify human baby food and as dietary supplements.

This naturalized Eurasian implant is also a pollen source for bees. It is high in vitamins A, B, C, E, and K and flavonoids, as well as calcium, potassium, iron, and protein.

It has been reported that alfalfa was brought to this country to increase yields of milk in dairy cows. In *Medicinal Plants of the Mountain West*, Michael Moore points out that because of modern agricultural practices, commercial varieties of alfalfa no longer accumulate as many nutrients, so its reputation as a milk increaser has diminished. He states that wild varieties are often more nutritious.

A genetically modified alfalfa has been grown in the United States. Because of concerns about safety and cross contamination, it has been embroiled in legal battles for several years.

RECIPE

Dried Alfalfa Supplement

Harvest flowers and leaves from top half of plant by stripping them off the stalk. Dry on trays, in bags, or by hanging in a dark place with good airflow. Once fully dried, place in an airtight jar. After a few days, open the jar, shake the contents, and reseal. If the moisture has redistributed itself and the plant does not seem fully dry, leave lid off for 6–12 hours and then reseal. Repeat every day until flowers are fully dried.

Can be stored whole or crushed into a powder. When ready to use, add to soups, stews, sauces, or smoothies for a nutritional boost.

Can also be used to make tea. Boil water and steep dried alfalfa greens for 5–10 minutes. Add honey and 1 teaspoon fresh ginger root if desired.

Alfalfa is a nitrogen fixer and so a great addition to a home garden. It may even find its way to your garden all on its own.

Range and Habitat

Grows in disturbed areas and pastures. In our region, it is most common in the foothills and mountains from 3,000'–9,000' in elevation. Typically grows in fairly nutrient-rich soil, which helps it accumulate many nutrients.

Comments

Flowers, leaves, and young shoots can be eaten raw, cooked, or dried. Generally, harvest only the upper half of the plant. Also, grazing of alfalfa makes it more bitter, so ungrazed plants are best. The seeds can be eaten sprouted or roasted or made into flour.

RECIPE

French Lentils and Alfalfa Flower Heads

French lentils are a gorgeous, smoky green or gray color and combined with the striking purple alfalfa flowers make a deeply stunning dish. Set your table to highlight these natural colors, or add a pop of complementary color with bright flowers or colorful cloth napkins.

Rinse 1 cup lentils in lukewarm or cool water until no foam appears. In a soup pot, bring the lentils and 4 cups water to a high simmer. Reduce immediately to a very low simmer. Cook uncovered for 20 to 30 minutes until lentils are soft. Make sure there is enough water that the lentils remain covered throughout the cooking time. If water becomes too low, add more. Do not allow to boil, as this will produce soggy, waterlogged lentils rather than distinct, firm ones.

Place ⅓ cup alfalfa flower heads into a medium-size bowl. When lentils are done cooking, drain water and immediately pour hot lentils over the flower heads. Allow the hot lentils to sit on top of the flower heads for a few minutes. Meanwhile, add ¼ cup raw chopped chives or wild onion, ¼ teaspoon salt, ¼ teaspoon pepper, and 1 teaspoon olive oil or butter. Mix gently so that you don't crush the flower heads. Serve warm or cold.

CLOVER, RED
Trifolium pratense

Family: Fabaceae
Look-alikes: White clover, alsike clover, alfalfa, golden banner
WARNING: There are many **toxic members** of the pea family. Make sure you positively identify this species before ingesting.

Description

This nonnative biennial or perennial has tiny, slender, pealike flowers that form dense balls at the top of thin stalks. The inflorescences are reddish, pink, or dark pink and often have a lighter-colored base.

Red clover grows about 6"–30" tall. The stalks are hollow, erect, and often somewhat sprawling. Leaves and stalk are hairy. Leaves are trifoliate, divided

> RECIPE
>
> **Fresh Clover Salad**
>
> Combine a handful of lettuce, a handful of arugula leaves, and a handful of clover (flowers and leaves). Toss together in a large bowl.
>
> In a separate small bowl, prepare the dressing. Combine 1 teaspoon olive oil and 1 teaspoon sweet rice vinegar. Add 1 tablespoon chopped green onions. Mix well until oils and liquids are combined. Pour over salad. Toss well.

into three leaflets. The leaves are ½"–1" long and alternate. They are green with a pale-green or whitish V-shaped marking (chevron). The white marking distinguishes red clover from alsike clover, which has only green leaves. Four to six branches per stem.

RECIPE

Sautéed Clover Flowers with Candied Pecans

Harvest 1 cup flower heads of red clover. In a heavy skillet, heat 1 tablespoon coconut oil to medium. Toss in ½ cup raw, unsalted pecans. Toss continuously until they start to smell roasted. Just as they begin to smell good and turn brown, decrease heat to very low. Drizzle 1 tablespoon honey over them. Add a pinch of salt. Toss for a minute or two until honey coats the nuts. Remove from heat, and transfer into a bowl.

Using the same skillet (unless you have burned the honey), bring 1 teaspoon olive oil or butter to medium heat. Add the clover flowers. Toss gently and sauté for 4 minutes. Add 2 cups chopped spinach. Toss together and sauté for 4 more minutes. Remove from heat and top with the candied pecans. Serve warm or at room temperature.

Range and Habitat

Red clover is a Eurasian immigrant that is now found widely across the United States. Especially enjoys disturbed ground and full sun. Will grow larger in moist areas. Found to 8,500' elevation or higher. Common garden weed.

Comments

This is the family of legumes, peas, and beans. Like them, clovers are nitrogen fixers and, therefore, splendid at improving poor soil—good for your garden and good for you.

RECIPE

Red Clover Tea

Use a teapot with a fitted steeper. Combine in the steeper 5 red clover flowers, 4 sprigs lemon balm from your garden, and half that amount of fennel leaves or root. When water boils, pour over herbs and allow it to steep about 8 minutes. Add honey and stir. Serve hot or cold.

   Variation: Add 1 clove garlic, fresh sliced ginger, or a pinch of turmeric.

CLOVER, SWEET
Melilotus officinalis

Family: Fabaceae

Other names: Common sweet clover, yellow sweet clover, white sweet clover, sweet clover, melilot, *M. albus*

Look-alikes: Alfalfa, all clovers, golden banner (poisonous)

WARNING: A mold that grows on sweet clover produces coumarin/coumadin. Coumarin is generally thought of as a blood thinner, but it actually prevents blood from clotting by inhibiting vitamin K production in the body of humans, cattle, and rodents. Vitamin K is an antidote to an overdose of coumarin. Coumarin is used to make rat poison. **DO NOT EAT MOLDY LEAVES.** They can be deadly.

Do *not* confuse with golden banner, which is poisonous.

Description

This sweetly pungent, very common annual, biennial, or perennial grows 1'–6' tall in disturbed areas. It's three-leaved like other clovers but taller. Alternate leaves are oblong or elliptic. Before it flowers, sweet clover looks similar to alfalfa.

Yellow or white flowers are small and pealike. Individual plants are either yellow or white flowered. Each flower is about ¼" long, forming into elongated,

> **RECIPE**
>
> **Sweet Clover Tea**
>
> Snip off enough flower stalks and leaves to fill your tea steeper. Boil water and then allow to cool for a few minutes. Pour over plant pieces, and steep for 3 to 10 minutes. Add honey if desired. On the trail, you can make sun tea by placing the ingredients in your water bottle and allowing the sun to do the rest as you hike along.

slender clusters or long, narrow spikes along branches, which branch off the tall main stalk. Blooms early summer to fall. Very strong taproot is impossible to pull out of the ground by hand.

Range and Habitat

This plant is a widespread invasive weed and often significantly hampers native species growth. It thrives in dry conditions but grows even bigger along streambeds. Harvest freely in such situations.

Comments

Flowers, shoots, and leaves are tender and edible raw or cooked. Some accounts say cooked roots are edible.

There is conflicting information about whether to use dried leaves due to the propensity for coumarin-producing mold to grow on the dried leaves. Other sources say dried leaves are edible and can be used to flavor pastries and soups. Use young leaves cooked or raw as you would any green. Shoots can be cooked similar to asparagus.

Personally, I stick to eating only the very fresh flowers or making tea with them. Fresh flowers are abundant and delicious: strong, pleasant tasting, and somewhat vanilla flavored. Flowers are used to flavor Gruyère cheese and can be dried and used as flavoring for soups.

In wetter years cattle have been known to die from foraging on sweet clover because of coumarin. This can also cause humans and animals to bleed to death even from minor wounds!

Many sources say that the plant itself contains coumarin; however, other sources explain that the production of coumarin is produced by mold.

Before eating this plant, get to know its life cycle. Be sure you can identify a fresh, non-moldy plant and distinguish between a withering, dying, or moldy one. If the plant does not look 100% fresh, I would avoid it. If you do ingest a moldy plant, consult your doctor immediately.

Sweet clover adds nitrogen to the soil and provides excellent nectar for bees.

RECIPE

Grilled Chicken with Yellow Sweet Clover Garnish

Marinate 2 chicken breasts in enough olive oil to coat them, ½ teaspoon salt, and ½ teaspoon pepper for 1 or more hours. Grill one side over medium heat for 5 to 10 minutes, then flip. Grill the other side for about 5 to 10 minutes until the chicken is no longer pink in the middle. Place on bed of lettuce and garnish with 2 teaspoons yellow sweet clover flowers.

CLOVER, WHITE
Trifolium repens

Family: Fabaceae
Other names: Dutch clover
WARNING: There are many **toxic members** of the pea family. Make sure you positively identify this species before ingesting.

Description
White clover is low growing, 4"–10" high. Its small, pealike flowers form a loose ball, sometimes with a pink tinge. These flower tufts are about ½" wide. Leaves are trifoliate (three-lobed), sometimes with a pale-green or whitish V-shaped marking.

FORAGER NOTE: Sometimes leaves are marked with white chevrons, sometimes not. Look closely at the individual flowers, looking for tiny purple spots and green teeth.

The species name *repens* means "creeping" in Latin, and true to its name, white clover creeps along the ground, planting itself in the soil (much like strawberry shoots) at its joints.

Range and Habitat

A Eurasian immigrant that is now found widely across the United States. Especially likes disturbed ground; cool, moist areas; and full sun up to the timberline.

> ### RECIPE
>
> **Peach and Clover Salad**
>
> Peach season on Colorado's western slope is short and very sweet. For an unusual local salad that is just as good for dessert as it is for an appetizer, harvest 1 cup clover flowers. Slice 4 fresh peaches into smaller than bite-size chunks. Combine in a bowl with the clover flowers. Add 2 tablespoons champagne vinaigrette; toss gently. Garnish with fresh sprigs of mint.
>
> **Variation:** Slice about 2 cups seedless red grapes in half and add to mixture.

Edible and Useful Plants

Comments

Flowers, stalks, leaves, and seeds can be eaten cooked, dried, or raw. White clovers are widespread and common; they make a great trail snack or fresh addition to a meal while camping.

Some people find that eating too many raw clovers causes bloating. To improve digestibility, clovers can be boiled with a bit of salt.

White clovers, like other clovers, are often found as garden and lawn volunteers (weeds). Clovers are a perfect example of how to turn weeding your garden into a fruitful and nutritious harvest.

Clovers add nitrogen to the soil and provide food for bees, other pollinators, and cattle. High in protein, they are a good forage crop for livestock.

Make a great addition to tea and salads. Wine can also be made from clover flowers.

RECIPE

Dried Clover and Rice Breakfast

Harvest white clover flower heads after they have gone to seed. Lay out on a tray and allow to dry in a dark room. Store in a sealed glass jar until ready to use.

Prepare short-grain brown rice according to rice's cooking instructions. When rice is almost finished cooking, prepare 1 fried egg. Heat skillet to medium high. Add butter or oil, and fry egg to your liking.

Fill bowl ⅔ full with warm rice. Place the egg on top. Sprinkle crushed dried clover seeds and flower heads on top, about ½ teaspoon. Sprinkle with Bragg Liquid Aminos and a bit of olive or flax oil. Add a sprinkle of hot pepper flakes if desired.

LICORICE, WILD
Glycyrrhiza lepidota

Family: Fabaceae

Other names: American licorice, sweet root

Look-alikes: Several species of sweet vetches (*Hedysarum* spp., which are also called licorice root), clovers, alfalfa, St. John's wort, Jacob's ladder, milk vetch (*Astragalus* spp.), and golden banner (**poisonous**)

WARNING: Do not ingest if you're pregnant or taking steroid therapy. Some sources warn that large amounts can be toxic and should be avoided by those suffering from liver disease, glaucoma, high blood pressure, or heart disease.

Edible and Useful Plants

Description

This aromatic native perennial can reach more than 3' high but on average is closer to 2' tall. It has taproots that can reach 3'–4' down and spreading rhizomatous root systems.

Leaves are pinnately divided and consist of organized rows of eleven to nineteen small, lance-shaped, opposite leaflets, each about ¾"–1½" long. There is always an odd number of leaflets, with a single one at the very tip of each leaf stem. Leaves can be a bit sticky. The leaf stem is often in a sweeping downward arch.

Flowers are similar to those of white clover (*Trifolium repens*), but the cluster is more elongated than the rounder clover flower head. Pealike flowers are white, yellowish, or greenish white (sometimes purplish). Flowers grow from leafless stems that originate at the axils (corners) between the leaf stem and stalk.

Brownish, spiny, burred seedpods are unique to this species within the pea family.

> **RECIPE**
>
> **Roasted Wild Licorice Root**
>
> Build a small fire in a safe area away from brush, shrubs, and low-hanging tree branches. Allow a large pile of coals to form. If you have a fire grate for grilling, place that over the fire and allow fire to burn down into coals.
>
> Place roots on the grilling grate. As each side just begins to brown, rotate and continue rotating numerous times for about 1 hour (depending on how hot the coals and how large the roots are). When the roots are soft but not burned, remove from heat and enjoy.
>
> *Variations:* Cut in half and sprinkle wild flaxseeds or evening primrose seeds. You can also bake the roots in the oven as you would a baked potato.

Range and Habitat

From about the Mississippi River westward to California and Texas and north to British Columbia. Moist fields, ditches, meadows, slow-moving streams, and other moist areas in the plains and foothills.

Comments

Leaves, young shoots, and roots are edible. All parts can be eaten raw or cooked.

Roots are sweet and often likened to sweet potatoes. Can be pounded to remove fibers before eating. Taproots are used for a variety of medicinal purposes. Taproots can be carefully dug out of the ground and dried in the sun. Harvest roots in the fall with a shovel. Roots can be 3'–4' deep. Roots also make a good tea.

Related to the commercial licorice variety (*G. glabra*).

FAGACEAE / BEECH FAMILY

GAMBEL OAK
Quercus gambelii

Family: Fagaceae
Other names: Scrub oak, Rocky Mountain white oak, Utah white oak
WARNING: Contains tannic acid, which can be poisonous if great quantities are consumed. Cattle will become ill if more than 50% of their diet is from oak. Frost increases toxicity.

Description
Common, native shrub oak with green, deeply lobed, waxy, alternate leaves. Seven to eleven rounded lobes per leaf, but exact leaf shape is variable. Scraggly shrub or small tree up to 75' high but often significantly shorter, more like 10'–35' tall. Moisture availability has an impact on height. Foliage often does not turn red like that of its eastern relatives.

Male catkins produce pollen and pollinate the female flowers; each produces one acorn. Acorns are about ¾" long and are held by a cap about one-third the size of the nut. Other acorn species have different-size caps; some even totally encapsulate the seed (acorn).

The complex root system consists of lignotubers (bud-producing swellings at the base of the stalk). They also have deep-feeding roots and rhizomes and sprout clones to form thickets of shrub oak. Also reproduces by seed, but vegetative spreading is more successful.

Range and Habitat
Wyoming and Nevada to Colorado and New Mexico and sparsely in Texas; also South Dakota. Dry slopes and canyons in the foothills and high deserts from about 3,300'–9,900' in elevation.

Comments
Acorns should be leached first and then can be eaten boiled or roasted. Acorns can also be ground into a flour.

Acorns should be leached or soaked before eating to remove the bitter and astringent tannins and to deactivate the antinutrient, phytic acid, to make the minerals more bioavailable and improve digestibility.

First remove the cap (shell or cupule) using a stone or hammer.

Acorns can then be leached through running water over several days, or by boiling. For the running water method, acorns can be leached by placing in a mesh bag and securing in a creek or you can devise a home method to mimic

Edible and Useful Plants

> ### RECIPE
>
> **Acorn Porridge**
>
> Combine ⅓ cup cold-leached acorn flour with ⅔ cups corn flour and 4 cups water. Add ¼ teaspoon salt, orange peel, and cinnamon. Bring to a simmer. Stir continuously until mixture thickens. Transfer to a baking dish and bake at 300°F for about 1 hour. Adjust water for desired thickness.
>
> **Variation:** Pour thick milk over the top prior to baking. Can be served as a sweet or savory dish. Dress it up however you like.

this. Acorns can also be leached by placing them or the flour in a large bowl of water and changing the water every few hours or at least twice a day, over several days. If using the flour method, then strain through cheesecloth or other fabric.

Cold-water leaching is generally preferred because the flour will retain its starch and can stick together in future recipes, whereas hot-water leaching denatures the starch and baked goods will crumble. There are countless recipes from around the world utilizing acorns. Many specify whether to use acorns that were hot- or cold-leached as results are different with the two methods.

Some claim that Gambel acorns can be eaten raw or without leaching, though others argue that leaching must be done to deactivate the antinutrients, even if the acorn tastes palatable prior to leaching. The best recommendation is to leach or soak no matter how it tastes raw.

Acorns can be made into porridge, bread, crackers, crepes, noodles, cake, and dips like acorn hummus. Acorns are an abundant wild food and a staple for many ancient civilizations. Researching recipes from around the world is quite fun, with endless recipe ideas from around the world.

Bark of the Gambel oak can be used for medicinal teas. Raw acorns can be stored in the freezer.

Be sure to harvest only fresh acorns without signs of mold, streaks of black, insect holes, or other signs of fungus or infestation.

A very important food for a wide range of wildlife, including deer, elk, bighorn sheep, squirrels, owls, and birds.

GROSSULARIACEAE / CURRANT FAMILY

CURRANT
Ribes spp.

Family: Grossulariaceae

Look-alikes: Thimbleberry and Boulder raspberry (larger leaves of similar shapes), Rocky Mountain maple (no berries), prickly currants (very similar but with lots of thorns), gooseberries

Related species: Northern black currant (*R. hudsonianum*), wax currant (*R. cereum*), golden currant (*R. aureum*)

Description
Currants are a native, deciduous, smooth-barked shrub ranging 1'–6' tall. Branches are erect and arching.

> FORAGER NOTE: The dried flower withers and hangs from the berry. This is perfectly fine to eat, but many prefer to remove it, especially when eating currants raw. Although the flower pulls off easily, it is small and time-consuming to remove.

Small, tubular, white, pinkish, or greenish, elongated, five-sepaled clusters of two to eight flowers give way to clusters of mild-flavored berries in late summer or fall. In drought years the process can speed up, and berries can be seen by midsummer.

Currant leaves can take on two distinct forms. Some are fan shaped and rounded, with shallow, rounded lobes (wax currant). Others are more deeply,

> RECIPE
>
> **Hearty Wild Currant Pancakes**
>
> Place your favorite organic multigrain pancake mix in a bowl. Add a pinch of fresh-ground nutmeg and 1/8 teaspoon fresh-ground cinnamon. Add water as directed. Stir gently.
>
> Heat skillet to medium-high. Add generous amounts of butter, coconut oil, or bacon grease, and heat. Spoon pancake batter onto skillet, and immediately add 1 teaspoon currants to each pancake. Flip when air bubbles begin to pop. Serve with real maple syrup or unbleached, fair-trade sugar.
>
> **Variation:** Add chopped almonds and flaxseeds.

sharply lobed similar to a maple leaf (golden currant). Leaves are alternate or in clusters.

Berries and leaves have resin glands that can make the berries appear dusty. Currant fruits are bright red, red-orange, golden yellow, or black, and they often appear somewhat translucent.

The golden currant (*R. aureum*) is named for its golden flowers. The fruit is a smooth berry that is green when immature. It ripens from yellow to red to black or dark purple.

Range and Habitat

Dry and sunny hillsides, forest edges, ridges, sagebrush fields, and disturbed areas from British Columbia to northern Texas. Hardy bushes are prolific; they literally grow out of rocks. They can be found thriving in dry and rocky soil throughout the region from about 5,000'–13,000' in elevation. Bushes grow larger with moisture.

RECIPE

Lemon Pudding with Fresh Wild Currants

Heat water in double boiler. Also heat 2 cups water in a separate pot to a boil. While water is heating, sift together 3 tablespoons cornstarch, 1 tablespoon flour, 1¾ cups granulated sugar. Whisk in the 2 cups boiling water until mixture becomes smooth. When water in bottom level of double boiler is simmering, add the flour-water mixture to the top dry pot; cook for 5 minutes, stirring occasionally.

In a separate bowl, beat 4 egg yolks. Stir several tablespoons of the hot mixture into the beaten egg yolks (to temper), then slowly add the beaten egg yolks to the hot mixture; continue simmering in the double boiler. Stir in grated rind of 2 lemons, 1 tablespoon butter, ¼ teaspoon salt, and ¼ cup lemon juice. Stirring constantly, cook until the mixture becomes smooth and thick.

Spoon into bowls, and top with fresh currants. Serve at any temperature.

Variation: Add a dollop of meringue or whipped cream.

Comments
Berries, flowers, and young leaves are edible. Can be eaten raw or cooked.

One fall afternoon while taking a roadside break with our bicycles, I found myself standing on a high ridge in the foothills outside of Boulder, Colorado. I was surrounded by a rolling field dotted with 4'-high currant bushes, all filled with bright reddish-orange berries. The aspen trees were turning yellow, and I could see for miles in all directions and up to the Continental Divide, where snow had recently dusted the high peaks. I smelled a whiff of sage and realized that fall was settling into the Rocky Mountains. This was currant season.

As I slowly picked wild currants from the great expanse, I thought about how many of these delicate little berries the local bears must pick to nourish their massive bodies in preparation for their winter slumber. This is slow-going work, but I get a sense that it has been done before, right here in this very field, and not just by human hands.

CURRANT, PRICKLY
Ribes lacustre

Family: Grossulariaceae
Other names: Bristly black currant
Related species: Mountain prickly currant (*R. montigenum*)
WARNING: Getting pricked by the spines can cause allergic reactions in some people.

Description
Very similar to nonprickly currant species. This native shrub is rambling and leggy, 1½'–5' tall, and covered in obvious sharp spines or thorns. Flowers are pink or coral.

Berries of *R. lacustre* are black. Bristly black currant is a native perennial that grows to about 3'–4' in height. Berries of *R. montigenum* are red.

Range and Habitat
R. lacustre has a larger geographic range than *R. montigenum*. It is found from Alaska east throughout Canada and south through Colorado and California. Found in shrublands, clear-cut areas, the subalpine zone, woods, and riparian woodlands.

R. montigenum is found throughout the western half of the United States and British Columbia. Found in alpine and subalpine zones.

Comments
Berries are eaten by bears, birds, and rodents. Use in the same ways you would use regular currants.

LAMIACEAE / MINT FAMILY

GIANT HYSSOP
Agastache urticifolia

Family: Lamiaceae
Other names: Nettle-leaf giant hyssop, horse mint
Related species: *A. foeniculum* (which is also commonly called blue hyssop), anise hyssop, lavender hyssop, licorice mint

Description
This strong-smelling perennial subshrub looks like really huge mint but with a scent more like a deep, perhaps burnt anise, or minty licorice aroma. Flowers cluster at the tops of stalks.

Giant hyssop (*A. urticifolia*): Pinkish, whitish, or lavender flowers bloom in late spring. Slowly spreads by rhizomatous roots. Stalks are single or branched

> **RECIPE**
>
> **Cold *Agastache* Gazpacho**
>
> Excellent in late summer when tomatoes are farm fresh and ripe.
>
> In a food processor, combine 6 large tomatoes and about 9 large hyssop leaves. Add hot or sweet peppers, such as ½ roasted Thai pepper and ½ roasted serrano pepper, although any peppers can be used. (To prepare peppers, roast on grill or in toaster oven until skin is browned. Allow to cool. Peel and remove seeds.) A roasted green bell pepper works well too.
>
> Add a splash of sherry vinegar, a splash of olive oil, and a dash of salt. Add a teaspoon of honey. Pulse until combined and well chopped but not totally liquefied and uniform.
>
> Serve room temperature or chilled. Garnish with fresh or dried parsley leaves and chopped chives or red onion.

and grow up to 3'–6' tall. Flowers are trumpet shaped with long protruding stamens and stigmas. They form in a tubular, gently pointed cluster along the top portion of the stalk 1'–6' long. From afar, appears somewhat similar to a clover head but elongated. Simple, pointed ovate (or widely lance-shaped) leaves are opposite, toothed, and widely spaced along the stalk. They are darker on the upper surface and lighter underneath. *(pictured)*

Blue/anise hyssop (*A. foeniculum*): Blue or lavender flowers with erect, sometimes branched stalks that reach 1'–4' in height. Inflorescences are clusters of small, trumpet-shaped flowers, together forming a cluster about 8" long at top of stalk. Coarsely toothed, opposite leaves, darker on top. Smells like licorice or anise. *(not pictured)*

Range and Habitat

A. foeniculum is found along the West Coast, comes into the Rockies, and is spotty in Colorado. It is relegated to the West, from British Columbia to California and

> **RECIPE**
>
> **Stewed Anise Plums and Peaches**
>
> Harvest 1 cup fresh flowers of the anise hyssop. Heat 6 cups water to a boil. Pour over the flowers, and let steep for about 8 hours.
>
> Optional: Add ¼ cup brandy. If desired, sweeten with 1 to 3 teaspoons honey or sugar. Add ¼ teaspoon cinnamon and 1/8 teaspoon nutmeg. Add a dash of vanilla, and stir ingredients together.
>
> Cut 3 fresh peaches and 3 plums in half and remove the pits. Place fruit halves flat side down in a saucepan that is large enough that the fruit is snuggled together but not crushed. Pour the tea mixture over the fruit.
>
> Bring to a low simmer over low heat. Now and again make sure that fruit is not sticking to the bottom of the pan by scraping under fruit with a spatula. If sticking, reduce heat. Simmer, covered, until liquid has mostly evaporated and has become thicker. Serve alone, over ice cream, or with a dollop of sweetened ricotta cheese. (Mix some sugar into the ricotta, and blend well.)

some parts of Colorado, but not New Mexico and Arizona. Favors meadows, foothills, creek banks, and roadsides.

A. urticifolia is found across the northern part of the country and from the Northwest Territories to western Colorado.

Comments

Seeds, leaves, and flowers can be eaten raw or cooked. The seeds of blue hyssop (also called anise hyssop) are best and can be used as a flavoring similar to anise or licorice seeds. Seeds can be used in sweet or savory dishes and added to jam or preserves. The anise-scented leaves are somewhat sweet tasting and can be used as tea and in potpourri. Nettle-leaf giant hyssop can be used in the same ways, but it does not impart the licorice-like flavor.

Well liked by bees, hummingbirds, and butterflies.

MINT
Mentha spp.

Family: Lamiaceae
Other names: Wild mint, field mint, poleo mint. Field mint (*M. arvensis*) is a common species in our area.
Look-alikes: Stinging nettle, giant hyssop
WARNING: Some people are allergic to mint.

Description
Wild mint has a square stem and smells like mint.

Wild mints are perennial, and many species are native. There are several species of wild mints, and they can smell similar to peppermint, spearmint, or some combination. Leaves vary but are basically ovate, mildly to very pointed, and toothed. Leaves are close to the main stalk and more or less horizontal to the ground. Leaves are spaced in opposite pairs along the branches and main stem. Rounded purple or lavender flower clusters grow either at the top of the stalk or in whorls along the stems at the corners where the leaves or petioles meet the stalk, depending on the species. Grows 6"–3' high.

Range and Habitat

Moderate to low elevations from Alaska to Florida. Especially in moist areas like banks of beaver ponds, fields, stream banks, and open meadows, but once established, it is widely tolerant of variations. Also found in plains, foothills, and meadows. Part shade or sun.

RECIPE

Fresh Mint Sorbet

In a saucepan heat 6 cups water just to a simmer. Add 1½ cups packed, finely chopped mint leaves. Stir and simmer for about 20 minutes. Remove from heat and add ¾ cup honey. Stir until dissolved. Place in refrigerator until fully cold.

Pour into prechilled ice-cream maker according to manufacturer's directions. Serve with a fresh sprig of mint.

Variation: Top with shavings of dark chocolate and a few drippings of honey.

Comments

There are several species of wild mints. Flavor varies, but all can be used interchangeably. Eat raw, cooked, or dried. Great for tea and flavoring.

RECIPE

Fruit Salad with Fresh Mint

Combine whatever fresh fruits are in season in a large bowl. Some to consider are peaches, plums, watermelon, and cantaloupe. Cut into bite-size pieces, 1 to 3 cups of each fruit.

Wash and pat dry ¼ to ½ cup mint, depending on how much fruit you are using. Chop roughly. Add to fruit mixture; toss to combine. Add fresh-squeezed lime or lemon juice if desired.

Serve immediately, or refrigerate overnight and serve cold.

RECIPE

Mint Tea

This is one of my staples throughout the growing season. It can be hot and dry in Colorado, so it's always good to carry a water bottle with you. I like to spruce up my water with a fresh sprig of mint.

Pick some wild mint and rinse off to remove dirt. Place sprig in water bottle, and let sit until you are ready to drink and enjoy fresh minty water. Over the course of the day, you can refill the bottle several times; the water will still become flavored.

LILIACEAE / LILY FAMILY

NODDING ONION
Allium cernuum

Family: Liliaceae
Other names: Wild onion, lady's leek
Look-alikes: Death camas (*Zigadenus venenosus*; **poisonous**), western blue flag (*Iris missouriensis*; **poisonous**)
Related species: Short-styled onion (*A. brevistylum*), taper tip onion (*A. acuminatum*), Geyer's onion (*A. geyeri*), prairie onion (*A. textile*)
WARNING: All wild onions can be confused with the toxic death camas. The basal leaf clusters look similar, but death camas flowers are quite different. They are conical-pyramidal clusters of cream or lavender flowers that are more tightly clustered than the onion flower heads. To be safe, identify wild onion for consumption when flowers are present.

Leaves and young shoots also look like the wild iris, called western blue flag (*Iris missouriensis*). This species looks like a garden-variety iris but smaller. It grows in similar habitats as wild onion—moist, open meadows—so again, it's best to identify onion when flowers are present, though onion grows in dryer areas as well where it is less likely to find the iris. Iris is **NOT EDIBLE**.

Description
This slender, delicate-looking native perennial has six to ten long, narrow, grass-like basal leaves that soar upward to about 1' tall and ¼" wide around the leafless stalk. Leaves are generally shorter than the flower stalk. Nodding onion grows

Edible and Useful Plants

> **RECIPE**
>
> **Sautéed Shrimp and Onion with Brown Rice**
>
> Cook 1 cup short-grain brown rice in rice cooker. Chop enough onion to make 2 cups of large pieces. Slice 1½ tablespoons fresh ginger into strips.
>
> Heat olive oil in a heavy skillet to medium-high heat. Sauté onion and ginger about 5 minutes. Add 10 shrimp. Sauté about 8 minutes, or until shrimp is done. Remove from heat and place in a serving bowl.
>
> Warm skillet again to medium and toss in 2 cups roughly chopped mizuna, chard, or other hardy green. Stir until wilted, about 4 minutes.
>
> Place brown rice in bowl. Add shrimp and greens. Top with flaxseeds. Serve hot.

to about 1½' tall but is often smaller. It has an oniony scent. One flower stalk emerges from each underground bulb.

Flowers are pink, light pink, or white. Flower heads are loose clusters of three petals and three sepals with long, yellow, anther-topped white stamens. Blooms mid- to late summer.

Nodding onion is notable by the way the flower head and the top of the peduncle (main flower stem) nod downward in an elegantly rounded sweep. Also, each flower in the cluster sits atop a nodding stem that also rounds downward. Seeds are black.

Range and Habitat

From British Columbia to Texas and spottily around the country. Found in full or partial sun, open meadows, roadsides, from the plains to subalpine zone throughout the Rockies.

Comments

All wild onions are edible. The bulb can be used like any onion—raw, dried, pickled, or cooked. Flowers and young shoots can also be eaten raw or cooked. Harvest is restricted in Arizona and some other states, so be sure to check your local listings before harvesting.

> FORAGER NOTE: Death camas does not smell like onion, so use your sense of smell in identification. Wild onions smell distinctly like onions. However, beware that because onions are so pungent, if there are onions and death camas growing in the same location, your nose can be fooled.

> ### RECIPE
>
> **Potato Crusted Onion Frittata**
>
> Preheat oven to 375°F. Coat the bottom of a 9" x 13" glass baking dish with olive oil. Slice several potatoes about ¼" thick and line baking dish with one layer of potatoes. Some overlap is fine.
>
> Add to baking dish on top of the potatoes: a layer of sliced tomatoes and a layer of broccoli or chopped kale, or a combination.
>
> In a mixing bowl, combine 4 farm-fresh eggs, 2 tablespoons homemade almond milk (or other milk), and ½ cup wild onion leaves, chopped. Add plenty of salt and pepper.
>
> Pour egg mixture over the vegetables. Bake until just firm and beginning to brown, about 40 to 60 minutes. With such dramatic variations in altitude in our region, baking times will vary. Adjust as needed.
>
> **Variation:** Add grated cheese to the egg mixture; beat in well.

Be aware that harvesting the bulb kills the plant. Harvest bulbs only when there is a large cluster of onions growing together. The stem, leaves, and flowers are plenty flavorful, though, so it's really not necessary to dig the bulb. Might as well leave it for next year.

I often see nodding onion growing singly or in small clusters. In this case, you can perhaps harvest a few leaves and a flower, but that is all that is acceptable.

LINACEAE / FLAX FAMILY

WESTERN BLUE FLAX
Linum lewisii

Family: Linaceae
Other names: Wild blue flax, blue flax, prairie flax, Lewis flax, *L. perenne*
Look-alikes: Chicory
WARNING: Green seeds should not be eaten; eat only the mature dark seeds. Some sources warn that wild flaxseeds **must be cooked** to remove cyanide compounds and that toxic reactions can occur if seeds are not properly prepared. Others believe they are perfectly safe to eat raw.

Description
This pretty native perennial flowers in spring and produces small edible flaxseeds in late summer. Its erect, branching stems are 3"–36" tall. Small (up to 1⅕") leaves are alternate, linear, or lanceolate and closely hug the stem. Basal leaves can remain green year-round. Does not flower until third (sometimes second) year. Light-purplish-blue flowers (rarely white) have five petals and are up to 2" wide.

Range and Habitat
Seen widely along roadsides; blooms in spring and early summer, going to seed by late summer. Ranges from Alaska to Texas in montane areas, roadsides, meadows, high mountain grasslands, and open fields up to about 9,100' in elevation. Early stage pioneer species, so often found in disturbed sites throughout the range.

Comments

Wild flaxseeds are a little bit smaller than store-bought varieties. Seeds can be eaten raw, roasted, dried, or sprouted. But see warning above.

I put flaxseeds on just about everything from rice dishes to smoothies to salads, and I often include them in baked goods like cookies. They are somewhat gelatinous and help hold baked goods together in much the same way that eggs do. Eaten raw, they are pleasant and smooth.

Harvest seeds in early fall. They fall out easily when the seed capsule is cracked and gently shaken. Seeds can be stored raw. Seeds can be eaten whole or ground. You can sprout them first and then eat them whole or ground. Many people prefer to eat flaxseeds ground rather than whole to better assimilate nutrients into the body. It is best to store flaxseed whole, however, to preserve oil quality.

This plant is related to common flax (*L. usitatissimum*), a European annual species that is cultivated for its seeds.

Pollinated by flies, bees, and other insects.

High in omega-3 fatty acids and fiber. Gelatinous, so it makes a good gelling agent for recipes, lotions, and hair gel.

Strong fibers can be made into cord, baskets, nets, and webbing for snowshoes.

> ### RECIPE
>
> **Raw Pumpkin Pie with Flax and Nut Crust**
>
> Place 1 cup raw almonds or cashews (or a combination) and 1 cup fat, juicy raisins or dates (or a combination) in a food processor fitted with the metal blade. Blend briefly; add ¼ cup flaxseeds and ½ teaspoon cinnamon. Blend very well, until all ingredients are finely chopped and begin to stick together like dough. Remove from food processor and press evenly into a pie plate.
>
> Fill with raw pumpkin pie filling or raw fruit filling of your choice.

MALVACEAE / MALLOW FAMILY

MALLOW
Malva neglecta

Family: Malvaceae
Other names: Common mallow, cheeseweed, cheese plant, button weed
Related species: Round-leaf mallow (*M. pusilla*), dwarf mallow (*M. rotundifolia*)

Description
This nonnative species of mallow can be an annual, biennial, or perennial. It reseeds readily and is difficult to eradicate once it takes over a garden.

Mallow is decumbent (lying flat along the ground with occasional erect stems), with one or many branched stems and sprawling arms up to 3' long, depending on growing conditions. Its herbaceous stalks are up to ¼" thick. Leaves are palmate, alternate, and about 2½"–3" wide. The leaves are roundish or kidney shaped with five shallow lobes, toothed, and wavy. Leaves have a deep indent at the base where they attach to the long petioles (leaf stems). They somewhat resemble small hollyhock leaves and are covered in tiny, hard-to-see hairs.

> **RECIPE**
>
> **Apple-Mallow Fruit Salad**
>
> Dice 1 apple into small 1" squares. Cover in fresh-squeezed juice of ½ lemon and toss to prevent browning. Toss with ½ cup mallow peas. Serve at room temperature.

Flowers are about ¾" wide, occur in groups of one to three, and are white, pinkish, violet, or white with pinkish stripes or markings.

The small fruits are round and light green, with a darker green-rimmed circle toward the center. These little buttons are sectioned and look similar to wheels of cheese. Fruits are ¼"–½" in diameter and are halfway surrounded by the calyx, which looks similar to a delicate, light-green leaf basket holding each fruit.

Range and Habitat

From British Columbia and throughout most of the United States, including Alaska. Mallow is a common garden weed found up to at least 8,900' in elevation.

> **RECIPE**
>
> **Mallow Gumbo**
>
> Mallow is so much like okra, it just screams to be used in a traditional New Orleans–style gumbo, especially since I almost never see any decent-looking okra being sold in stores in the Rocky Mountains (although it can be grown in the warmer parts of the Rockies quite successfully). Substitute young mallow leaves for okra in equal quantities in any gumbo recipe. The fruits can also be used, but I have a hard time preventing myself from eating them raw to save a sufficient quantity for anything else.

Comments

Leaves and young shoots are edible raw or cooked. The young tender leaves are especially good in salad and smoothies. The little fruits, or peas (cheese wheels), are delicately mild and delicious.

To harvest the buttons, pluck the fruit from its calyx basket and eat as is, in a salad, or cooked. They are smooth and melt in your mouth. Flowers are edible raw but should be harvested and eaten soon after or they will wilt. Mallow flower tea is also good. As mallow seeds age, they turn brown and can be sprouted and eaten like any sprout.

Soak any part of the plant in water and the water will become gooey and mucilaginous. Use the gooey water as lotion. Add essential oils for extra home-spa yumminess.

Can be soaked in water and the thick water can be used to make marshmallows and a meringue pie topping.

> **RECIPE**
>
> **Raw Mallow Wheels**
>
> Harvest the fruits and remove the leafy calyx that partially surrounds them. Serve raw in a small bowl with other light appetizers such as olives, cheese, apple wedges, and crackers.

ONAGRACEAE / EVENING PRIMROSE FAMILY

EVENING PRIMROSE
Oenothera biennis

Family: Onagraceae
Other names: Common evening primrose
Look-alikes: Great mullein
WARNING: Can have a sedative effect for some. Take caution when driving, caring for children, and so on, until you know how this plant affects you.

Description
Native biennial that forms a starlike rosette in its first year and sends up a stalk 2'–8' tall in spring of the second year. Leaves are narrow and linear, with a prominent light-green or whitish-colored midvein. Leaves are alternate along a stalk about 6" long.

Thick, sturdy, erect stalks occasionally branch once or twice.

Yellow, four-petaled flowers form tight clusters along the top portion of the stalk. Open in afternoon and evening. Fruits are hairy, tubular seedpods that point upward from the stalk for a short time with the remains of the flowers sticking out of their tops.

Roots are whitish; their size depends on how compact the soil is and how much moisture was present.

Range and Habitat

British Columbia to Texas. Moderately dry roadsides and fields up to about 9,000' in elevation.

Comments

Roots are eaten raw or cooked. Can be roasted, fried, boiled, dried, or sautéed. Some people advise boiling roots more than an hour and changing water at least once to improve flavor.

Flowers, flower buds, and leaves can be eaten raw or cooked. Roots are best harvested from the first-year basal rosette during fall, winter, or early spring before the stalk emerges.

RECIPE

Boiled Root and Leaves with Winter Stew

Harvest roots of several basal rosettes after the first hard frost. Scrub well. Cut into large chunks about 2"–3" long. Bring large pot of salted water to a boil. Add primrose roots. Boil for about 15 minutes; drain and change water. Bring a new pot of salted water to a moderate boil, and boil roots again for 15 minutes; drain, rinse, and set aside.

In the now-empty large pot, heat olive oil. Add 1 cup roughly chopped onion and 5 cloves garlic, chopped. Sauté with 4 cups wild game or beef (stew cuts are fine) cut into 2"–3" chunks, until browned but not cooked through.

Add the boiled roots and 1 cup primrose leaves. Add a few cups chicken or veggie broth, enough to cover the meat and roots. Add 3 bay leaves and 2 carrots, roughly chopped. Bring to a boil, then reduce heat and simmer with a cracked lid for 30 to 60 minutes, stirring occasionally. Make sure the meat is cooked through.

Serve hot over rice, with fresh crusty bread or a side salad.

Raw seeds are high in linoleic acid and gamma linoleic acid and make a good omega-3 supplement.

NOTE: The hairy-throat feeling left by the leaves and roots can be unpleasant.

> ### RECIPE
>
> **Smoothie Topped with Primrose Seeds**
>
> This is one of my favorites, and come fall, I seem to constantly crave it. The recipe is very simple.
>
> Fill a blender container halfway with orange juice. Add 1 organic banana, 2 small, handpicked apples (seeds and core removed), 6 ice cubes, and a pinch each turmeric and ginger (or fresh ginger root). Blend well on high speed until ice is pulverized. Pour into pint glasses and top with a pinch of cinnamon and ½ teaspoonful of primrose seeds. Enjoy.

FIREWEED
Epilobium angustifolium

Family: Onagraceae
Other names: Common fireweed, perennial fireweed, narrow-leaved fireweed, great willow-herb, rosebay willow-herb, alpine fireweed, blooming sally, *Chamerion angustifolium*
Related species: Dwarf or broad-leaved fireweed (*E. latifolium*), a shorter relative

Description

Erect native perennial reaching 2'–8' tall with showy pink flowers. Smooth stems, mostly unbranched, green or sometimes reddish.

Leaves are narrow, hairless, lanceolate or linear, with smooth or lightly toothed margins and short petioles. They have a pronounced white midvein and grow to 2½"–5½" long. Leaves are long, narrow, and willowlike and very similar to the leaves of its relative, the evening primrose. Lots of alternate leaves along the stem form a spiral pattern.

Flowers form in a loose, elongated triangular raceme cluster along the top portion of the stalk. Very showy flowers are pink, magenta, or purplish (rarely white), each about 1" wide, with four petals and four sometimes-darker sepals. Flower clusters are about 8" long along the stalk and can bloom throughout summer. The entire raceme does not bloom at the same time, so it is common to see developing seedpods along with blooming flowers. Each flower produces 300 to 500 seeds, which travel by wind to reproduce.

Grows in dense clusters from hardy rhizomes that can extend 15"–17" deep into the soil and spread horizontally underground. Rhizomes survive after forest fires, making fireweed one of the early post-fire ecosystem recovery pioneers.

Range and Habitat

From Alaska across the West to New Mexico, all of Canada, and into the northeastern United States. Riparian areas in alpine, subalpine, and montane zones. Early pioneer after fires. Roadsides, pastures, riverbanks, burned forests, and rocky riverside outcrops.

RECIPE

Steamed Young Shoots

Make note where the fireweed stand is and go back in spring to harvest young shoots. Gently steam and top with olive oil or butter and salt, or toss lightly with homemade pesto.

Comments

Edible parts include young shoots (raw or cooked), older stems (peeled, raw, or cooked), leaves (raw, cooked, or dried and fermented to be used for tea), roots (raw or roasted), and flowers (raw or as jelly or candy). Can also be made into salves and balms.

Many accounts say the older leaves are too bitter to eat. They sometimes give a somewhat prickly-hairy mouth feel that is not pleasant.

Dry old stalks can also be used to make cordage.

Eaten by moose, deer, caribou, muskrats, hares, and other wildlife. Taking some cuttings early in spring can stimulate additional growth, but this species is very sensitive to being trampled, so care should be taken not to step on fireweed stands.

PINACEAE / PINE FAMILY

At this moment in time, the forests in the state of Colorado where I live are struggling pretty badly. I think it's a similar situation throughout the Rockies. In the late 1800s and early 1900s, the old-growth forests were decimated. We now have large stands of spindly trees packed too close together. The old-growth ecosystems are no longer in existence. The pressure of pine beetles and forests fires is an ongoing threat to forest ecosystem recovery. The small poorly established trees are far less resilient to both assaults and so we have many standing dead within our forests, which just makes them more susceptible to forest fires.

This section on trees is honestly a bit reductive. I have only provided snippets about the edible qualities of a few trees. I do not dive into the complexity and importance of old-growth forests; however, I wanted to at least mention it here. My hope is that our society will prioritize restoring old-growth ecosystems with the diversity needed to thrive in the era of climate change and beyond. We cannot have a healthy planet without immense stretches of old-growth forests.

As much as the mountains still beckon and are filled with happy recreationists and foragers, I do find a sadness to walking through toothpicks struggling to reclaim their rightful place in the world. I find a sadness in the denuded understories. I wish I lived in a time before industrialization where I could have experienced what a real forest had to offer. The best I can do now is talk about it, bring light and urgency to it, and leave a better world to future generations. I invite you to join me in the movement to restore old-growth forests across the country and planet. What a legacy we could have, standing tall and wide with our woody, rooted friends.

Let's learn a little bit about some trees of the Rockies.

BLUE SPRUCE
Picea pungens

Family: Pinaceae
Other names: Colorado blue spruce, silver spruce, *pino real, P. parryana, P. commutata*

Description
This stately, almost iridescent native conifer (cone-producing evergreen tree) reaches 70'–115' in height. Stiff, stout, prickly needles are often noticeably and dramatically blue hued. The color can range from green or turquoise to powdery or iridescent blue. New needles can be from powder blue to greenish. Needles are four-sided and about ½"–1¼" long with white stripes. Cones are long, narrow cylinders that are about 2½"–4½" long, light brown and papery, and notable by the wavy teeth along the scale tips. Shake hands with blue spruce and it will bite. Seed production begins after 20 years.

Range and Habitat
Idaho, Wyoming, and the Four Corners states; very limited in Montana. A fairly common mountainous species; found on high slopes, in mixed forests, and along stream banks from about 6,700'–11,500' in elevation. Does not grow all the way

up to tree line. Prefers moister areas. Often found in Douglas fir and Engelmann spruce forests and along waterways, often with cottonwoods.

Comments

The inner bark is nutritious and can be dried, ground, and mixed into other foods. It is not recommended to harvest bark from the trunk of a tree, as this can severely compromise the tree's ability to bring nutrients to the branches and can result in the tree's death. Instead, harvest inner bark from small side branches.

RECIPE

Blue Spruce Beer

Spruce tips can be boiled in water, then the water used to make fermented wine. The beer recipe below uses hops for a more classic beer flavor. To make blue spruce beer, you need the following:

1 large cooking kettle (8 quarts or larger)
Several 1- to 2-liter glass bottles with tight-fitting lids, such as beer growlers or 40-ounce beer bottles (2-liter soda bottles can also be used), cleaned and sanitized with sanitizing solution
Funnel
1¼ gallons fresh water
3–6 ounces fresh blue spruce tips, including needles and stems (fresh-growth tips in springtime work great, but any time of year is OK)
½–¾ ounce dried hops or 1–2 ounces fresh hops
1 small to medium-size piece of ginger root, bruised and cut into chunks
2–3 cups sugar or molasses, depending on desired sweetness
1 package baker's yeast

Bring the water to a boil; add the spruce tips, hops, and ginger. Continue with a light boil for 40 minutes, with a lid partially covering the kettle.

Remove from heat and pour the mixture through a strainer into another container. Add the sugar or molasses and stir until completely dissolved and mixed. Allow the mixture to cool until the pot is just warm to the touch.

Add the yeast, and stir vigorously for a couple minutes. Cover pot with good-fitting lid; let sit for 24 to 48 hours.

Using the funnel, pour the mixture into the bottles. Fill each container halfway at first. Shake it around to aerate the mixture, then fill the bottle completely, leaving about 1" of headspace. Secure the lid tightly and let the bottles sit for about 5 days at room temperature (around 70°F).

Tea can be made from the fresh needles, which can also be chewed to freshen breath. Some sources say that young male catkins can be eaten raw or cooked and that immature female cones, cooked or roasted, are sweet and gooey. The seeds are nutritious and high in fat. Needle beer is also good. Young shoots, stripped of their needles, can be eaten raw.

The tallest blue spruce ever recorded was 126' tall; the oldest was 600 years old. It is the state tree of Colorado and Utah, and it is used for shelter by moose, deer, owls, the Jemez Mountains salamander, and the northern goshawk, among others. Blue spruces are also used by bald eagles for breeding.

DOUGLAS FIR
Pseudotsuga menziesii

Family: Pinaceae
Other names: Rocky Mountain Douglas fir, Douglas spruce, Oregon pine
Look-alikes: All evergreen conifers

Description

This fragrant native conifer (evergreen) tree is not a true fir (*Abies* genus). In the Rocky Mountains, it grows to about 100' high. The trunk reaches about 3' in diameter.

Notable by the three-pronged bracts that extend from beneath the scales of the cones. The story goes that to escape a forest fire, lots of frightened mice scampered over to the Douglas fir trees and took shelter under the shingles of their cones. When you look at the cones with a bit of imagination, you can still see the little feet and butts of the mice sticking out.

Cones are light brown and 2"–4" long. They hang from the branches and are readily

> FORAGER NOTE: Look for the mouse feet and tail sticking out of the cone.

Edible and Useful Plants   215

visible both on the tree and on the ground below. (Compare to true firs, which have cones only toward the very top of the trees, and they disintegrate before falling to the ground.)

The needles are flat, not sharp and prickly, and have white lines on the underside. They are dark green, bluish, or yellow-green and ¾"–1¼" long. Bark changes with age.

> ### RECIPE
>
> **Marrow and Ground Cones**
>
> Grind cones well in food processor or blender until powdered.
>
> Combine 1 cup powdered Douglas fir cones in saucepan with 2 cups bone marrow. Cook over low-medium heat until combined and marrow has liquefied. Cook at a very low simmer for a few more minutes. Allow to cool.
>
> Serve in a small spoonful as a garnish. This delicacy goes well atop dried bite-size pieces of summer squash or fresh crusty bread.

> ### RECIPE
>
> **Needle Tea**
>
> Pick enough needles from the tree to half-fill your teapot steeper. Place in teapot and pour just-boiled water over the steeper. Put lid on while steeping to retain the aromatic qualities; steep about 10 minutes. Enjoy alone or with raw local honey.
>
> Needles are available year-round, so this is a great source of vitamin C and a healing astringent in the dead of winter. No need to plan ahead.

Range and Habitat

Coastal and mountainous tree found from British Columbia and Alberta to New Mexico from sea level to 11,000' in elevation. Common in montane and subalpine zones and forested hills.

Douglas fir is a very common tree in the Rocky Mountain region. It also grows in coastal areas, where it is much larger and reaches 80'–200' or more in height and 15' across.

Comments

Inner bark can be eaten. It is best to harvest inner bark from smaller branches rather than the trunk. Harvesting from the trunk can kill or severely compromise the health of the tree, especially if not done properly. If harvesting from the trunk, take only a small portion, about the size of your hand, and never girdle the tree (i.e., do not take a strip that goes all the way around the trunk).

As with all conifers, you can enjoy chewing on the raw needles. They are soft and delicious. Use as a breath freshener or in place of chewing gum or mints. They also make a great tea and can be used to make a simple syrup for flavoring drinks and desserts.

Douglas fir is the second-tallest tree in the United States, having been recorded up to 329' tall. (Only the redwood is taller.)

PINE, PIÑON
Pinus edulis

Family: Pinaceae
Other names: Colorado pinyon, 2-needled piñon, pinyon pine, Rocky Mountain pinyon pine

Description
This relatively small, compact, bushy native evergreen tree often has a crooked trunk and spreading crown. It grows 15'–45' tall. Needles are green, less than 1" to about $4\frac{1}{3}$" long and remain on the tree for about 9 years each. Needles are usually in clusters of two but sometimes one or three.

Cones are rounded, about 2" long, and somewhat bulbous with very thick scales. They ripen from green, purplish, or reddish to brown. Each female cone produces ten to twenty delectable seeds, two behind each scale. Seed production begins at about 25 years of age, and best seed production is at around 75 years.

Range and Habitat
Found along the rim of the Grand Canyon and throughout the foothills and deserts of the Four Corners region from about 4,600'–8,900' in elevation. Common

> **RECIPE**
>
> **Pine Nut Trail Mix**
>
> Pine nuts can be used raw or lightly toasted.
>
> In a bowl, combine ½ cup pine nuts and any of the following: raisins, dried currants, walnuts, dried apples, dried cherries, granola, and chocolate chips. Bring on trail hikes for an energizing snack.
>
> **Variation:** Combine 1 teaspoon honey and 1 tablespoon melted ghee or coconut oil, and mix very well. Stir into fruit and nut mixture until mixture is lightly coated with the oil-and-honey mixture. Spread onto a cookie sheet and bake at low heat (around 225°F) until just browning on the edges. You can also use a food dehydrator. Remove and let cool. Break into pieces and pack as a trail snack.

in dry areas, piñon-juniper scrublands, and regions with hot summers and cold winters.

Comments

Pine nuts are a true wild delicacy with their rich flavor and deep nutrient profile. Trees produce heavy seed crops about every 3 to 7 years. Seed production is better in more moist years. Reproduction of this species occurs only by seeds, which are so delicious they are often eaten by wildlife.

A typical tree can live 500 years, some for 1,000 years. A small piñon pine shrub with only a 6"-diameter trunk can be hundreds of years old.

Many of the piñon pines still standing throughout the Southwest are the very same trees that provided sustenance to the pre-Columbian people who lived in the Four Corners area—the Navajo, Hopi, Anasazi, and their ancestors and neighbors.

Ponder that with me for a moment, will you? A massive change has happened in the world, and the trees that provided this nutritious delicacy are still standing, not

just in history books but in real life, right here. It changes the meaning and the romance of history when we realize that that old world still exists. If we are willing to step away from our phones for long enough, we too can experience what the world is really meant to be and who we are meant to be in it.

In the harsh, dry desert of the southern reaches of the Rocky Mountain range, great and sophisticated societies of well-fed people, of hunters, of cultivators, of foragers, of medicine specialists, the children of many tribes and descendants of more than 10,000 years on this continent, picked the nuts from these very trees. I can just imagine the great piñon harvests.

This rich, fatty food should be cherished and appreciated along with the slow, deliberate nature of the great Southwestern desert.

RECIPE

Classic Dairy-Free Basil Pesto

Blend well in food processor 1 cup pine nuts, 4 cups packed fresh basil, 6 cloves fresh garlic, ¾ cup olive oil, and 1/8 teaspoon each of salt and pepper. Serve on toast, dehydrated zucchini slices, or pasta.

Variations: Substitute fresh mint for basil. Add 1 fresh jalapeño.

PINE, PONDEROSA
Pinus ponderosa

Family: Pinaceae
Other names: Big heavy, black jack, bull pine, ponderosa white pine, Sierra brown bark pine, silver pine, western pitch pine, western red pine, western yellow pine, yellow pine, Yosemite pine
WARNING: Some sources warn that needles can harm unborn babies and are reported to be an abortifacient. **Pregnant women should avoid.** Pregnant cows and other mammals will lose their calves if they consume too many pine needles. Large or frequent use of ponderosa pine needles or pitch can cause kidney problems in some people.

Description
Native to the western half of the United States, this large evergreen tree grows 100'–180' tall (sometimes to 250', making it one of the world's tallest pine trees). Its diameter is 2'–6' wide.

The trunk is straight and usually branchless. The trunk is pungently vanilla smelling. The bark is rough and cinnamon or orange-brown colored with black or gray, charred-looking markings.

Edible and Useful Plants

> ### RECIPE
>
> **Ponderosa Needle Tea**
>
> Harvest a half-handful or so of needles. Place in tea steeper with 2 sprigs mint and 5 rose hips. Pour boiling water directly over tea mixture. Allow to steep for 5 to 20 minutes. Add honey, stir, and enjoy.
>
> **NOTE:** This makes an excellent holiday gift for friends and relatives. Place ingredients together in a decorative jar.
>
> **CAUTION:** Do not use during pregnancy.

Needles are long, about 5"–11", and usually grouped in threes, though not always. The needle clusters are surrounded at their base by a small papery sheath. Together, the needle clusters grow in spacious, rounded tufts. Trees have male and female cones. Female cones are notable by the wide, rigid, thornlike pricker on the underside of each cone scale. Seeds mature in late summer to fall of the cone's second year.

Range and Habitat

Found from southern British Columbia to New Mexico and throughout the West from sea level to about 9,800' in elevation. Often found in mixed stands with Douglas fir, lodgepole pine, juniper, blue spruce, and quaking aspens, among other species.

Comments

The seeds, inner bark, young male cones, pollen, resin (pitch), and needles (with limitations) are edible.

Collect the inner bark anytime, but it is best in spring when the sap is flowing. It is described as tasting like sheep fat. To collect inner bark (of any tree), never take it from the trunk, as this can seriously damage or kill the tree. Harvest from side branches, and think about proper pruning techniques as you do so.

If you must harvest from the trunk, do *not* ring (girdle) the tree unless your intention is to kill the tree (forest thinning for forest health or fire mitigation purposes). Rather, slice small vertical rectangles. Look for the white inner bark and avoid eating the outer bark. Take a strip that is no more than one-tenth the circumference of the trunk. It can be eaten raw but can cause stomach cramps.

It is better cooked. Inner bark can be roasted, boiled, fried, or dried and pounded into flour and used in baking or stews.

Young male cones can be eaten cooked. Harvest before they open and boil. Small oil-rich seeds can be harvested from female cones in late summer or fall of the cone's second year. They can be eaten raw or cooked.

Needles are used medicinally in tea for colds, coughs, fevers, and other medicinal purposes. **Not for pregnant women.** Warmed pitch can be used to help remove splinters.

RECIPE

Fried Cambium

Harvest inner bark by stabbing tree with a sharp knife. Cut a small square from the tree. Remove outer bark and shave off strips of cambium layer. It will be white or cream colored. Cut into thin strips like french fries. Fry in hot fat or oil and top with sea salt.

PLANTAGINACEAE / PLANTAIN FAMILY

COMMON PLANTAIN
Plantago major

Family: Plantaginaceae
Other names: Ribwort, broadleaf plantain, buckhorn plantain, rippleseed plantain, greater plantain, white man's footprint
WARNING: Do not confuse plantain leaves with those of young hellebore, which is poisonous.

Description
Simple, roundish green leaves are low-growing basal rosettes and notable by the prominent ribs on their undersides. Leaves grow to about 6" long and 4" wide and lie more or less horizontally along the ground (sometimes they pop up a bit).

This nonnative perennial herb produces a light-brown, leafless flower stalk 3"–16" high. Tiny yellowish or greenish flowers tightly hug the stalk and bloom throughout summer and into fall. Seeds ripen from midsummer to fall. All parts are mucilaginous.

Range and Habitat

From Alaska to Texas and across the United States in disturbed areas, lawns, and along roadsides. Found in the plains and foothills throughout the Rockies in both dry and moist areas. A common plant found around the planet, including gardens, banks of ponds, and clearings.

> FORAGER NOTE: This plant shares a common name with the tropical banana-like fruit, also called plantain. They are not related and do not look alike at all.

Comments

Leaves, stalks, and seeds are edible. Leaves can be eaten as any leafy green vegetable. Young leaves, and those not exposed to harsh conditions such as excessive drought, are more tender. It can help palatability to remove the very fibrous midvein and other ribs from the leaf. Tougher leaves can be boiled.

Seeds can be eaten raw or cooked (often boiled). Young flower stalks can also be eaten like green beans.

These small plants often go unnoticed as just another common weed. I have heard that chewing plantain leaves can be effective in creating an aversion to smoking tobacco. Use as a cold-infused tea or poultice. Seeds have fiber and can be used as a supplement and a laxative or, like psyllium, for intestinal health.

Plantain is astringent, antimicrobial, and anti-inflammatory. Leaves can be used as a poultice for insect bites and poison ivy.

RECIPE

Poultice for Insect Bites

Crush several clean leaves by chewing them gently between the teeth. Place directly onto welt caused by an insect bite. Hold in place until pain and swelling subside. Depending on the severity, you may want to replace with a fresh poultice after a while. If you are treating someone else, you may choose to crush the leaves in your hands to make the poultice or have the patient chew the leaves themself. **NOTE:** This is not a remedy for poison-infused bites, such as those from a venomous spider or snake.

POLYGONACEAE / BUCKWHEAT FAMILY

BISTORT
Bistorta bistortoides

Family: Polygonaceae
Other names: *Polygonum bistortoides*, American bistort, western bistort, bottlebrush, miner's toes, knotweed, smartweed, snakeweed (because of its snake-like root system)
Look-alikes: Buckwheat, clover
WARNING: Harvest minimally and thoughtfully in delicate alpine ecosystems. I have seen warnings indicating that excessive consumption of bistort may cause photosensitivity.

Description
This erect, native perennial grows about 8"–24" tall, and it is taller than most plants in similar ecological zones, so it can be easy to notice among the shorter high-alpine foliage.

The inflorescence is a terminal raceme, or cluster of flowers, that forms a white or light-pink tuft at the top of each stand-alone stalk. Each cluster of flowers is about 1"–2.4" long and less than an inch (0.3"–0.8") in diameter.

Tiny stamens extend beyond the corolla and protrude past the petals (stamen are exserted). Some sources report that the smell is unpleasant.

Thin, lance-shaped basal leaves grow up to about 6"–10" long. A few smaller, thin, alternate leaves creep up the slender green or red-green stalk.

Range and Habitat
Found in montane and alpine zones from British Columbia and Alberta down to California, and east to Colorado and New Mexico. Found in high alpine, moist meadows and slopes, and near streams.

Comments
Leaves, flowers, seeds, and roots are edible.

Raw bistort roots are generally bitter and astringent due to the high tannin content. Best to cook the starchy roots before eating. Roots can be roasted, baked, or boiled. Roasted roots are said to taste somewhat similar to chestnuts.

As with most wild greens, young leaves are better than older leaves for eating raw and have a citrusy flavor. Use to flavor soups and stir fries.

Bistort seeds can be eaten raw or cooked. To eat raw, add to salads as you would chia or flaxseeds. A bistort flour can be made by dry-roasting the seeds.

Allow to dry, and then grind into a flour for use in baking or add to stew or soup for extra nourishment.

High-alpine meadows are delicate habitats. Please be mindful when tromping through the meadows. Harvest only if bistort are abundant, and even then, leave plenty to reproduce.

I absolutely love very old recipes, and if you do too, check out recipes for bistort pudding. It's fatty and savory and sounds amazing, though I have not yet tried it.

> ### RECIPE
>
> **Roasted Bistort Root**
>
> I always hesitate to share bistort root recipes because of the potential for overharvesting and destroying a stand of bistort, so please harvest thoughtfully.
>
> Place cleaned bistort roots in a pie dish or baking sheet and bake at 400°F until tender. Time will vary depending on how thick the roots are and your altitude. Slice open and first try some plain so you can intimately experience the bistort root. If you like, add butter or olive oil and sea salt. Particularly delicious with roasted lamb or jerk chicken and a fresh side salad.

CURLY DOCK
Rumex crispus

Family: Polygonaceae
Other names: Narrowleaf dock, sour dock, yellow dock (because the root is yellow), crispy dock, dockweed
Related species: Willow dock (*R. salicifolius* or *R. triangulivalvis*), western dock (*R. aquaticus* or *R. occidentalis*), broadleaf dock (*R. obtusifolius*), canaigre dock (*R. hymenosepalus*), pale dock (*R. altissimus*)
WARNING: Contains oxalates. See oxalates discussion in the introduction.

Description
Docks, perennial Eurasian implants, have low basal leaf clusters that grow to about 1' across. Erect, sometimes branched clusters of flower stalks stick straight up. Flowers are without petals. They are light green or greenish red, growing in tight, erect, whorled clusters. The flowers give way to burnt-red or dark-brown clusters of smooth, three-sided achenes with a large seed lump visible, often lighter colored, in the center. Inflorescences are about 6"–18" long.

Curly dock grows 1'–5' tall. Stalks are round, hairless, and ridged. It is notable by its large single, curly, wavy, or undulating leaf margins, hence the name. Leaves are entire, lanceolate, and hairless. Basal leaves are the largest. Leaves are smaller, with shorter petioles along the stalk. The light-green or purplish midrib is pronounced. The root of curly dock is yellow.

Curly dock is distinguished from golden dock by its seedpods. Golden dock seedpods are serrated and have long needlelike points sticking out. It also has leaves coming out from between the seed heads, whereas curly dock leaves do not.

RECIPE

Stuffed Zucchini with Dock

Summer is almost synonymous with creative zucchini recipes. Here's one of my favorites, and it is amenable to endless variation.

First, cook 1 cup sticky rice. Leftover rice works great too.

Preheat oven to 350°F. Slice 2 large zucchinis in half and scoop out the seeds. This leaves canoe-shaped zucchini halves. Lightly coat the bottom of a glass baking dish with olive oil. Use a dish big enough to fit the 4 zucchini canoes. Place canoes in baking dish skin side down.

Take ½ cup young dock leaves. Wash and pat dry. Chop leaves into very small pieces. Add chopped veggies including ¼ cup green onions, 2 carrots. Add hot pepper and salt to taste. Chop all together into tiny pieces. Combine with the dock and rice. Add 2 farm-fresh eggs. Mash it all together.

Add veggie-rice mixture to inside of zucchini boats. Drizzle with olive oil and sprinkle with paprika. Bake until everything is soft, browning, and bubbling.

RECIPE

Dock Leaf Nachos

Start with a stack of 6 fresh organic corn tortillas. Heat 2 tablespoons oil in a large skillet to medium-high heat. Cook tortillas in pan until they're soft, one side, then the next. Remove tortillas from pan and crack into several small pieces. Place chips into a glass pie dish.

Preheat oven to broil.

Evenly spread 1 cup black or pinto beans across the top of the chips. Add 1 cup finely chopped young dock leaves and 3 tablespoons chopped onion. Add ½ cup chopped green pepper. Add black olives. Top with cilantro and chopped jalapeño. Add cheese if you like.

Place in oven on second-highest rack in center of broiler coil. Rotate after 2 minutes. If using cheese, cook until cheese is bubbling and just begins to brown. Remove from oven and serve as is or with fresh tomatoes or salsa.

Variation: Use premade corn chips. If dock leaves are tough, boil them first for 7 minutes; drain and add to nachos as described above.

Range and Habitat

Commonly found throughout the plains and foothills regions, especially in ditches and disturbed ground from Alaska to Texas. Curly dock and golden dock are the more common docks found in Colorado.

Docks are seen commonly along roads, ditches, and gravelly roadsides throughout the region. The erect, burnt-red (sometimes yellow or light-green) flower or seed heads are easily visible as you drive by.

Comments

Leaves, young flower stalks, and petioles are edible raw or cooked. Seeds are also edible but bitter. Leaves are most tender and least bitter when young. Stalk can be peeled to reveal the sweeter inside portion. The leaves taste somewhat like kale with a note of sour. They lose their toughness when cooked. Leaves are good simmered in soup stock or stew.

Roots and young leaves are used medicinally for liver issues.

FORAGER NOTE: Even though the flowers are green, it is more common to see the burnt-red seedpods standing erect along roadsides.

Seeds can be used as flour, to make piecrust and crackers, or to flavor vinegar. They must be rubbed to release the buckwheat-shaped, three-sided seed from its casing. Seeds are rich in iron but bitter. Some reports say leaching improves taste.

The *Rumex* genus consists of about 200 species, including docks and sorrels. In photos, dock and sorrel look very similar, but in real life, it is fairly easy to tell them apart. Docks are much bigger, with tougher leaves. Dock leaves themselves, however, can vary quite a bit. Dock species are known to hybridize with one another, making it somewhat difficult to distinguish between specific species. All are edible in the same ways, so it is not essential to identify the specific species.

SORREL, ALPINE
Oxyria digyna

Family: Polygonaceae
Other names: Alpine mountain sorrel, mountain sorrel
Look-alikes: Sheep sorrel (*Rumex acetosella*) has similar flowers and seeds but different leaves. Also, alpine sheep sorrel/few-leaved dock (*Rumex paucifolius*).
WARNING: See oxalates discussion in the introduction.

Description

This native perennial looks similar to the docks but is much smaller. It is similar to sheep sorrel in size but with different leaves and fatter, fleshier flowers and seedpods.

Fleshy leaves are somewhat fattish and round or kidney shaped. They have a pronounced indent where the petiole meets the leaf. They grow in clusters of low-lying leaves with erect protruding flower stalks. Flowers grow in dense clusters along the stalk and

> FORAGER NOTE:
> Mountain sorrel is distinguished from the docks and sheep sorrel by its rounded or kidney-shaped leaves.

Edible and Useful Plants

are green or reddish, becoming darker red as they become winged achenes, or seedpods.

Range and Habitat
Western United States from Alaska and the Northwest Territories to California and Texas. Common in the higher mountains of the West. Found in alpine, sub-alpine, arctic, and tundra zones, especially nestled in rocks; most often in moist areas near streams or melting snowpack.

Comments
Leaves, stems, and seeds are edible raw or cooked. Tart-flavored leaves can be used similar to rhubarb or to make a lemonade-like drink. Eat leaves and stems raw, cooked, or pickled as you would any versatile leafy green vegetable.

SORREL, SHEEP
Rumex acetosella

Family: Polygonaceae
Other names: Common sheep sorrel, red sorrel, sour weed, field sorrel, spinach dock, garden sorrel, *Acetosella vulgaris*
Look-alikes: Mountain sorrel, the docks
WARNING: See oxalates discussion in the introduction.

Description

This nonnative perennial is a *Rumex* similar to the docks, and the erect flower stalk that gives way to rust-colored achenes is similar to that of the much larger, sturdier docks. It is difficult to see the difference in photos, but sheep sorrel is much smaller and more tender than the docks.

Stems are very slender but erect, 4"–24" tall. Whale-shaped leaves are less than 1"–6" long and less than 1" wide.

Sheep sorrel has a characteristic simple (not compound) leaf shape that resembles a whale with wings, or fins, at the leaf base. Not all leaves, even on the same plant, will have these curlicues, but when they do, they are about one-quarter the size of the leaf or smaller. Leaves are otherwise oval or widely lanceolate.

Inflorescences are whorled flower spikes arranged in a narrow pillar along the stalk. Flowers are maroon, red, or light green in color. Fruits are achenes (small seedlike structures) and are rust colored. The herbaceous perennial blooms June to August.

Reproduces by seed and by creeping rhizomatous roots about 8"–2' (some reports say 5') underground.

Range and Habitat

Naturalized from Eurasia, sheep sorrel is now considered a noxious weed in about half of the US states. Grows from Alaska to Texas and across the United States with the possible exception of the Deep South. Sheep sorrel colonizes acidic and low-nitrogen (poor) soils, grass fields, pastures, disturbed sites, clear-cuts, and recent burn areas from about 4,000'–11,200' (lower in the northern areas) in elevation throughout the Rockies. Grows in moist, riparian areas, on dry open slopes, and along roadsides. Shade-tolerant, it is also found in forests. Found all the way down to sea level in other parts of the country. Hardy to USDA Zone 3.

> FORAGER NOTE: Much smaller than the docks. The leaves remain tender even after the plant has gone to seed.

RECIPE

Polish Sorrel Soup

Cook 4 eggs in boiling water for 8 minutes until hard-boiled. Cool by soaking in cold water, peel, and set aside. Cook 4 medium-size potatoes cut into quarters in boiling water until tender, about 15 minutes. Drain and set aside.

Bring 6 cups chicken or vegetable broth (homemade or bouillon) to a boil; reduce to a simmer. Add 1 cup chopped sorrel leaves and return broth to a simmer.

In a separate bowl, mix 3 tablespoons organic white flour with 12 ounces sour cream; stir into broth. Add salt if desired.

Divide cooked, quartered potatoes among four soup bowls. Place 1 hard-boiled egg, cut into quarters, into each bowl. Pour soup into prepared bowls. Serve hot.

Variation: Leave out the sour cream but still add the flour. In this case, sauté the flour in organic canola oil for 2 minutes on medium to low heat, stirring constantly. Then add it to the broth and stir in. The soup will appear creamy but without the dairy.

> ### RECIPE
>
> **Trail Recipe**
>
> Pack a blackened tofu or roast beef sandwich and add a big pile of sheep sorrel leaves that you find along the way. You'll love me for this one.

Comments

The first time I discovered sheep sorrel, I felt like I had discovered some old French culinary delicacy. It was delightful. The fact that sheep sorrel is classified as a noxious weed in twenty-three states (which means herbicide city—no more poisons, please) is mind-boggling. This is one of the loveliest wild edible greens I know of. Its taste is fresh, light, and with a unique lemony punch. The texture is so smooth it melts in your mouth. Sheep sorrel eradication efforts could be a great opportunity for foragers to start working with local governments and farmers. Offer to harvest this wonderful crop and save them the time and money of buying and applying more poison.

The leaves, roots, flowers (raw or cooked), and seeds (ground into flour) are edible. Leaves can be used as a curdling agent for milk products.

Sheep sorrel has a long history of cultivation around the world. It has been used in traditional recipes in Russia, Ukraine, Lithuania, Hungary, Turkey, Poland, Bulgaria, Nigeria, Greece, and other countries.

PORTULACACEAE / PURSLANE FAMILY

PURSLANE
Portulaca oleracea

Family: Portulacaceae
Other names: Common purslane, pursley, wild or common portulaca, pusley, pigweed (although amaranth is more commonly referred to as pigweed)
WARNING: Contains oxalates. See oxalates discussion in the introduction.

Description
This prostrate (flat-growing) succulent ground cover has smooth, thick, reddish stems and smooth leaves that are often nearly opposite. Leaves appear in whorls or clusters at the stem joints and tips. Leaves do not reach much more than 1" long but together form a thick and spreading ground cover. Small, solitary, yellow flowers bloom from stem ends but are very short-lived and rarely seen.

Purslane's taproot is fibrous and branching, making it well suited to drought conditions. Its tiny, black seeds form within a little nest, visible when the short-lived flower falls away.

Range and Habitat

Common garden weed from British Columbia through the Rocky Mountain region to Florida. Found in gardens, lawns, and disturbed areas. An old-world native, there is some debate about whether there was a pre-Columbian species in our hemisphere. Purslane is a hot-weather plant that sprouts in midsummer once the soil has warmed.

Comments

Leaves, stems, and seeds are edible raw, pickled (like sauerkraut or pickles), or cooked. Purslane can also be dried, but because it is fat and juicy, drying is not most people's preferred method.

Because purslane is so low-growing, it is usually quite covered in dirt. It is such a sturdy plant, though, that cleaning is easy. Soak in a few changes of water, disturbing it to release the soil, or rinse using a colander.

Purslane has a strong, sour-lemony taste and goes well with other flavorful foods. Spanish speakers of the Southwest called pickled purslane *chao chao*. It can be made with peppers, zucchini, and onions.

Purslane is high in omega-3 fatty acids. Purslane is also high in vitamin E and other nutrients. An important component of the Mediterranean diet, purslane has been an important staple and ceremonial food crop around the world for thousands of years. It is also a great companion plant for breaking up soil, as its roots dig deep into hard-packed ground, as well as ground cover to retain moisture. Think twice before pulling purslane as a weed. Instead, allow it to benefit your garden and provide essential fatty acids and other nutrients to your diet.

The uses for purslane are endless. Its gooey quality helps thicken soups and stews and can be used similarly to okra for this purpose, although the taste is more pronounced than okra. Purslane can also be used in drinks where thickness and a lemon flavor are desired.

RECIPE

Good Old-Fashioned Sausage, Onion, and Purslane

Grill 4 links spicy sausage and slice into bite-size pieces. Slice and then sauté 1 medium-to-large yellow onion in olive oil in a skillet. When onion is halfway to caramelized, add 1 cup chopped purslane. Continue to sauté until almost caramelized. Toss with sausage; serve on a bed of sushi rice. Garnish with garden-fresh parsley.

ROSACEAE / ROSE FAMILY

APPLE
Malus domestica

Family: Rosaceae
Other names: Orchard apple, table apple
WARNING: Apple seeds contain amygdalin, which degrades into cyanide in your digestive system. One apple's worth of seeds would not be expected to cause harm for an adult, but if eaten in large quantities, it might. Too much can cause respiratory failure or even death.

Description
One of the world's most-beloved fruits, there are more than 7,500 cultivated varieties of this species. Native to western Asia, where it was first cultivated thousands of years ago.

Small trees, sometimes rambling looking, with many branches and full leaves. Apple trees are somewhat squat and grow 14'–33' tall.

Simple, alternate leaves are serrated and 1"–4" long. Clusters of white, pink, or red five-petaled flowers. Apples vary in size and can be from red to green. Apples also vary in taste from sour to sweet, depending on the variety.

Range and Habitat

Highly cultivated temperate region fruit tree requires warm summers and does best with even moisture and not too much wind. Even though not as common, apple trees can grow in higher elevations and have been known to succeed up to 9,800' in elevation but usually lower. Sun best; partial shade OK.

Comments

Apples can be eaten raw, cooked, or dried. They can also be frozen. Fruit is high in pectin, which is good for making jams.

Apple trees were widely cultivated throughout the period that European immigrants were setting up homesteads in the western United States. Many of these orchards have now been cut down to make way for development. Even so, apple trees can still be found almost everywhere humans have settled.

I found an apple tree in an abandoned field one fall day and was consumed by thoughts about the person, the pioneer, the farmer who struck out to the unsettled West, to the unknown, and planted this tree. I wanted to thank the ancestor who planted this apple tree and tell him how it is doing.

It's very big and old now. Its bark is thickly ridged, and its lower branches droop nearly to the ground. They are dotted prolifically with apples. I like how they grow together in clusters of four, light reddish and striped and mottled. The tree is so tall and gnarled, and still it is amazingly fertile. Apples absolutely cover its crown, well above where I stand on the ground.

I walk inside its giant sweeping arms, and no one on the sidewalk can see me. I could pick your tree's apples in the privacy of its embrace. The bugs are sharing some of the apples, but not too many, except where they blanket the ground. I eat one while it is still attached to the branch and think about who you might have been. I know you were thoughtful and took care about what kind of apple tree you planted because these apples are the sweetest. I know you thought about it, about how to feed your family. I want you to know that you are feeding the future. And the future is grateful.

Update: After the first edition of this book was published, developers removed this poetic old tree along with the entire grove. An old field in Boulder, Colorado, no longer is home to such relics of a bygone past where food was grown among the people. Now we have buildings and concrete and oil wells, as well as memories of what once was, and what might be again with renewed focus on cultivating local food systems.

Recipes

An entire cookbook could be written about apples, but I'll narrow it down to just a few recipes.

> ### RECIPE
>
> **Dried Apples**
>
> I love this simple recipe. Having unadulterated, perfectly dried apples in the pantry through winter is a huge (and cheap and healthy) treat. This can be tasty even with the sourest apples.
>
> Leaving a square around the core, slice apples into quarters. If large, slice each quarter in half. Place in bowl with fresh-squeezed lime juice and toss. After slicing a few apples, place them in the bowl with the lime juice until finished. Place slices onto food dehydrator trays. Dehydrate until dry. Store in sealed glass jars.
>
> **Variation:** Skip the lime juice and put apples directly onto dehydrator trays. Great for holiday gifts.

> ### RECIPE
>
> **Frozen Apples**
>
> To freeze, slice apples and remove the cores. Spread slices out on cookie sheets and place in the freezer. After several hours, when slices of apples are frozen, remove from cookie sheet and place in large plastic food storage containers. Store sealed in freezer. Use throughout the winter to make pie, apple crisp, muffins, or smoothies.

> ### RECIPE
>
> **Baked Apples**
>
> Preheat oven to 350°F. Remove cores from several large apples. Spray a glass baking pan with coconut oil. Place apples in pan. Into the holes left when you removed the cores, add honey, walnuts, fresh-ground cinnamon, and fresh-ground nutmeg. Cook until apples are collapsing and honey mixture is bubbling. Serve warm.
>
> **Variation:** Serve with vanilla ice cream.

CHOKECHERRY
Prunus virginiana

Family: Rosaceae
Other names: Chokecherries, bitter-berry, wild black cherry, Virginia bird cherry, *Padus virginiana*
Look-alikes: Common buckthorn/European buckthorn (not edible), serviceberry (edible)
WARNING: Hydrocyanic acid causes the seed, bark, and leaves to be toxic. Seeds (which are like small plum pits) are poisonous. **Do not eat them.**

Description
This native perennial is a deciduous, thicket-forming shrub or small tree growing 6'–40' high. Sometimes takes the shape of a tree but often grows as thickets with thin trunks about 6" wide. It has extensive rhizomatic root systems and reproduces either by spreading rhizomes or by seed.

Chokecherry leaves are shiny, alternate, and oval or oblong, with tiny, jaggedly toothed edges. They are green, with the underside lighter than the top, turning yellow in fall. Similar to the leaves of other stone fruits, chokecherry leaves are 1"–4" long; the width is about half the length.

> ### RECIPE
>
> **Chokecherry Jelly**
>
> Wash cherries and remove stems. Simmer 4 cups chokecherries in enough water to cover the fruit. While fruit is simmering, mash it with a wooden spoon. Remove from heat and run cherries through a Foley mill, retaining the liquid and pulp in a bowl below.
>
> Put juice-pulp mixture back into a saucepan. Combine with 1 packet pectin. Bring to a boil, and then add 1 to 2 cups sugar. Stir well. Bring to a rolling boil. Continue boiling for 1 minute, stirring constantly. Remove from heat and place in jars. Store in refrigerator.
>
> **NOTE:** For long-term storage, use canning methods appropriate for jelly. Place in sterilized glass jars with sealed lids, as you would other jellies.
>
> **Variation:** Use half fresh apple cider and half chokecherry juice.

Small, nicely scented flowers are white and form together in showy, tubular, drooping clusters or racemes 3"–6" long. Flowers in spring to summer.

Fruits are available from late summer to early fall. Fruits are tart, dark drupes about ¼" in diameter. The berrylike fruits hang in elongated clusters mimicking the shape of the flower clusters.

Range and Habitat

From Alaska south across most of the United States. Absent from several of the southeastern states. Found commonly in the foothills and canyons of the Rocky Mountains, especially next to creeks and ditches.

Comments

Chokecherries are very common. The plant is named for its astringent, mouth-puckering quality; however, the edibility of raw chokecherries varies by

plant and location. While I find some cherries are not enjoyable to eat raw, many are delicious right off the branch. Chokecherry is commonly used to make jam, syrup, and wine. It goes well on a big slice of fresh toast or homemade pancakes.

Note that the seeds are very hard and will break a juicer. You are warned. A Foley food mill works better for removing pits from flesh.

Despite the warning that the seeds are poisonous, it is widely considered fine to boil them with the fruits before separating them out. Also, you can eat chokecherries raw and then spit out the seeds.

PLUM, AMERICAN
Prunus americana

Family: Rosaceae
Other names: Goose plum, river plum, wild plum

Description
This native perennial shrub or small tree grows 3'–33' tall. In the Rockies, it more frequently looks like a brambly shrub. Leaves are typical of the stone fruits—ovate, usually with pointed tips, sharp teeth, and reaching 3"–5" long. Can form thickets by underground-spreading root systems. Although individual shrubs are short-lived (20 years or so), the stand can live far longer.

Inflorescences are single or in clusters of two to four fragrant, showy, white flowers, about 1" each. Fruits are somewhat smaller than commercial plum varieties but still quite

plump and fleshy. They are yellowish, bright-rose, reddish, purplish, or orangish drupes or stone fruits. The seed is a typical pit (although smaller), as in commercial plum varieties. Ripens in late summer to early fall.

Range and Habitat

From Saskatchewan and Manitoba to New Mexico and Arizona and across portions of the eastern United States. Requires at least 16" of rainfall per year, so in the arid West is usually found along stream banks, ditches, or other moist

> ### RECIPE
>
> **Dried Wild Plums**
>
> Harvest a basketful of plums. Remove seeds if desired, but it's not at all necessary. Place on racks of food dehydrator. Dry until plums are dried but not hard as a rock.
>
> Store in a sealed jar through winter. Enjoy as a snack on their own or in baked goods or winter stews or compotes.

Edible and Useful Plants

> ### RECIPE
>
> **Grilled Plum and Goat Cheese Salad**
>
> Slice plums in half and remove pit. Prepare salad with mixed greens from the garden (baby kale, chard, dandelion greens, lamb's-quarter, and a few amaranth leaves). Combine organic olive oil with good balsamic vinegar and a touch of honey. Mix well and toss with salad greens.
>
> Grill or pan-fry halved plums over medium heat, turning to prevent sticking and burning.
>
> Serve in individual bowls with several grilled plums per person. Top with crumbled goat cheese and fresh-ground pepper.

riparian habitats or in meadows with moisture-capturing depressions. Often found in canyons mixed in with grapevines and chokecherries and along roadsides. Up to about 7,500' elevation in New Mexico (somewhat lower farther north).

Comments

Wild plums are for sure one of my favorite wild foods. Eat them raw, dried, stewed, in pie, grilled, as fruit leather, with muesli, or straight from the tree.

RASPBERRY
Rubus spp.

Family: Rosaceae
Other names: Wild red raspberry, American red raspberry. *R. idaeus* is the most common species in the Rocky Mountain region.
Look-alikes: Thimbleberry, black raspberry (*R. leucodermis*)

Description
Low-growing deciduous shrub with a perennial root system that sends up biennial stalks from which small, five-petaled flowers produce red raspberries in late summer or fall. The shrubs grow 1'–10' high but are usually seen throughout the Rockies closer to 3'–4' in height. Sharply pointed leaves are alternate and pinnately compound in leaflets of three to five.

Like its relative the rose, raspberry is also covered in thorns, so take care when picking. Fruits are actually aggregates of drupelets that are commonly referred to as berries. Berry size and production can vary widely depending on the plant and conditions. Wild raspberries are generally smaller than commercial varieties.

Range and Habitat

Raspberries are generally widespread throughout the United States. *R. idaeus* is also widespread but is not found in the southern states from Texas to Florida. Found in disturbed areas, sun-dappled openings in forests, and along roadways and woodland borders. Also found on steep slopes but usually smaller here. Does

> ### RECIPE
>
> **Old Fashioned Raspberry Pie**
>
> This is my Grandma Buddy's old-fashioned pie recipe, with a few variations. I learned it with a Crisco crust, but here is a slightly healthier variation.
>
> Preheat oven to 400°F.
>
> **Filling:**
> Measure proper amount of raspberries by filling your empty pie dish to overflowing. Transfer berries to large bowl. Add ¾ cup unbleached cane sugar, grated rind of ¼ lemon, and 4 tablespoons tapioca pearls. Stir gently to combine; set aside.
>
> **Crust:**
> Prepare chilled water by filling a bowl with water and placing a few ice cubes in the bowl. Set aside.
>
> Using a food processor with the metal blade, pulse to combine ingredients. Combine 2 cups white pastry flour and 1 teaspoon salt. Cut in ¾ cup butter until mixture resembles coarse crumbs with some small pea-size pieces. Sprinkle in 2 to 4 tablespoons chilled water, 1 tablespoon at a time, and pulse to combine. If needed, add 2 to 4 additional tablespoons of the chilled water and combine. Dough should be workable, not too wet and not too dry. Don't overmix.
>
> Shape dough into two even balls. Roll out one ball at a time, one for the top crust and one for the bottom crust. Roll dough on a lightly floured work surface into circles 2" wider than the pie plate. Transfer one section to pie plate, and gently press along the bottom.
>
> Fill unbaked pie crust with the raspberry mixture. Add a few slices of butter to the top of the heap if desired. Place the top crust on top of the berries; fold and pinch edges to secure top crust to bottom crust. Using a fork, poke a few holes in the top crust. Place pie in preheated 400°F oven. After 10 minutes, reduce heat to 350°F and bake for about 45 minutes to 1 hour, or until juice is actively bubbling out of the crust and the crust is just beginning to brown.
>
> **NOTE:** Place a cookie sheet under the pie plate to catch any juice overflow.

well in full sun and partial shade. Can tolerate both dry and moist conditions. Hardy to USDA Zone 3.

Comments

Raspberries are primarily pollinated by bees.

Berries, leaves, and roots are edible. Berries can be eaten raw or cooked. Leaves can be used fresh or dried as tea. Leaves are said to be useful for the female reproductive tract, uterine health, and alleviating menstrual cramps.

Leaves should be harvested when the leaf is still green. Make tea with fresh leaves or fully dry leaves. Some accounts say that partially dried leaves are not good for this use.

Raspberries are a great gateway wild edible because most of us already know what they look like and have spent our lives eating them. Learning to identify them in the wild is easier for this reason.

> ### RECIPE
>
> **Hot Cereal and Fresh Raspberries**
>
> An excellent addition to oatmeal or amaranth. Great while camping. Prepare hot cereal according to instructions for that cereal. Add a few fresh-picked wild raspberries. Top with a spoonful of crushed almonds or walnuts and local honey.

The psychological aspect of foraging cannot be overstated. Most people in our culture have no experience with picking and eating wild plants and have a huge mental barrier to doing so. Part of the hesitance is the fear of eating something poisonous. This is of course totally rational. Overcoming this fear is not just a mental exercise. It takes serious study of wild plants. Some species require more study than others.

With proper precautions, raspberries can't reasonably be mistaken for anything poisonous, and they are incredibly prevalent throughout the Rocky Mountain region. They grow along trails (sunny disturbed areas especially) throughout the entire United States, so the beginner has access to wild raspberries even on easy hiking trails.

ROSE, WILD
Rosa spp.

Family: Rosaceae
Related species: Common species include Woods' rose (*R. woodsii*), prickly rose (*R. acicularis*), and Nootka rose (*R. nutkana*).
WARNING: Some accounts warn to not eat the seeds because they contain cyanide compounds and have hairs that cause discomfort.

Description
This lovely native perennial shrub is spreading, can form dense thickets, and grows 1'–10' high (often around 3'–4'). Stems are thorny. Leaves are alternate, and each leaf consists of five to nine leaflets (pinnately divided) that look the same as cultivated rose leaves. Leaves have one oblong, serrate leaflet at the tip of the leaf stem, with one to four pairs of matching, opposite leaflets. Leaflets are ½"–2" long.

Beautifully rose-scented, five-petaled, light to dark pink flowers are 2"–4" wide and flower in spring to summer. The round central disc of pollinating parts is yellow.

Edible and Useful Plants   253

About fifteen to thirty-five seeds are encased in a fruit called a rose hip. Rose hips look somewhat similar to red currants but are thicker, meatier, and not as translucent. Like currants, the shriveled flower is often still hanging from the fruit. Rose hips can vary significantly in size and color, from big, shiny, and bright red to smaller, duller, or more subtly colored. They are about ¼"–½" long.

Range and Habitat

From Alaska to California and Texas. Considering the high price of commercial roses, you'd think wild roses would be rare, but they are not. Wild roses are found almost everywhere around our region. I see them prolifically along roadsides, stream banks, and trails from the plains to mid-elevations. They grow larger with moisture and loose soil but also grow in dry and more-compacted areas.

RECIPE

Fried Tofu with Sweet-Spicy Rose Hip Sauce

It seems like the hot peppers in my garden are ready at just about the same time as the rose hips. So why not try them out together? To make the sauce, combine ½ cup rose hips (fresh or dry) with 3 cups water; simmer for about 10 minutes. Remove from heat, and smash the rose hips so the seeds pop out. Simmer for 10 minutes more; remove from heat.

Strain liquid to remove the seeds. Return liquid to saucepan. Add a few hot peppers, diced. The amount varies depending on how spicy the pepper is and how spicy you want the sauce to be. Add 1 heaping tablespoon sugar or honey. Simmer for 10 minutes, stirring frequently.

Mix ½ teaspoon organic cornstarch with a little bit of water. Add to simmering mixture and stir in well. Sauce will thicken. Cover and turn off heat.

Variation: Add a dash of tamari, fish sauce, and ginger while simmering.

Slice extra firm tofu into large slices about ½" thick. Heat skillet on medium to high and add olive or coconut oil. Fry tofu on each side until browned and firm. Remove from heat.

Serve tofu on a bed of chopped lettuce and pour plenty of rose hip sauce over tofu. Sprinkle some chopped green onions on top.

Edible and Useful Plants

Comments
Rose hips (fruits) and flowers can be eaten raw, cooked, or dried. Flowers make a nice garnish for sweets and salads. Rose hips are used as a base for rose hip jam (*champe*). Fruits can also be used to make necklaces and garlands. Rose hips can be made into sauces both spicy and sweet. Rose hips can be eaten at any stage and are especially good after a frost sets the sugars.

Over one hundred species are in the *Rosa* genus, and many are cultivars. Several wild rose species are found in the Rockies; they hybridize with one another, so exact species identification can be difficult. Luckily, it is not necessary, as all rose species can be used interchangeably.

SERVICEBERRY
Amelanchier alnifolia

Family: Rosaceae
Other names: Saskatoon, juneberry, western serviceberry, alder-leaf shadbush, shadberry, shadblow, western juneberry, sarvisberry, pigeon berry
Look-alikes and related species: Blueberry, low serviceberry, common serviceberry
WARNING: Some sources warn that the seeds contain cyanide-like compounds and advise you to cook or dry before eating large quantities. Other sources say to eat as you would blueberries.

> **RECIPE**
>
> **Dried Serviceberries**
>
> It is wonderful to dry the berries in a food dehydrator and store for use throughout the winter. Add to pancakes, muffins, and smoothies or make a homemade serviceberry trail snack with your favorite nuts, seeds, and other dried fruit. Spread fresh berries in your food dehydrator trays and process on low until dry.

Description

This native, deciduous, perennial shrub or small tree is commonly seen from 3' to 15' tall but can reach heights of about 29'. If maturing to tree size, it can be large with spreading branches, but often it is seen shorter and in dense clusters as a thicket.

Small but conspicuous white flowers emerge in early spring. Each flower has five thin petals about three times as long as they are wide. Flowers are borne in terminal clusters, or racemes, of about five to fifteen flowers. Look for dense sprays of white blooms in spring turning to dark-purple edible berries as the season progresses. Each berry contains about four to ten seeds. The edible berries

> ### RECIPE
>
> **Simple Compote**
>
> Make a simple compote by simmering the berries and mashing them up a bit with a wooden spoon as they simmer.
>
> Start by placing fresh washed and destemmed berries into a saucepan. Add about ¼ cup water. Water amount varies depending on amount of berries and size of pan but does not have to be exact. Use enough water to have about an inch layer on the bottom of the saucepan.
>
> The berries will release water as they simmer so you just need to add enough water so they don't stick to the bottom until they release their juices. Don't worry about adding too much as you can always simmer a bit longer so that the moisture boils off if you want a thicker compote.
>
> As the mixture simmers gently, smash berries with a wooden spoon. Bring entire mixture to a simmer and cook for about 15 minutes or longer.
>
> Transfer to glass jars or plastic storage containers. If storing in the freezer, plastic is a great choice to avoid broken glass, but allow mixture to cool before transferring. Store in freezer and enjoy through winter.
>
> This is a great method because the compote does not require added sugar for preservation. You can pull it out of the deep freeze throughout the winter and use as you would applesauce. Eat it plain or use as a topping for your favorite meats or crackers, with melted cheese, or to top ice cream or crepes.
>
> **Variation:** Can be turned into serviceberry jam.

look quite similar to blueberries, with a crown on the end. They are dark blue or purple, sometimes dark red.

Simple, ovoid leaves are alternate with serrated margins (toothed edges) toward the top and smooth toward the petiole where the leaf attaches to the stem. Leaves are from less than an inch long to about 2" and display prominent, light-colored veins. They are sometimes hairy and sometimes smooth.

Bark of the shrub or tree is smooth and gray or purplish.

This plant has an extensive, massive root and rhizome system, and most new plants are shoots rather than seedlings.

Range and Habitat

A. alnifolia is the species native to the Rocky Mountain region, but similar varieties are found throughout the continent. Species hybridize easily, which can make species identification difficult.

Grows from just above sea level to tree line (about 160' above sea level in California to 10,000' in Colorado). Grows in open woods and hillsides from foothills to montane. Often found in habitat with chokecherries and hawthorn. The serviceberry needs sun and will die out as the canopy of a maturing forest blocks the sun. Often found on the edges of forests and roadsides. This species is found in a wide variety of ecosystems, but it avoids flood zones and is common soon after a fire or logging.

Comments

The succulent, delicious, deep-purple berries look very much similar to blueberries. The flavor is deeper, though, enhanced by the rich almond extract–like flavor. Can be dried in a food dehydrator, simmered to make compote, used to make jam, fruit pies, muffins, or use as you would blueberries. Can be eaten raw, dried, or cooked.

A single plant will live about 6 to 20 years, though an 85-year-old member of the species has been recorded.

Folklore says that the common name serviceberry refers to the time in early spring when both the flowers of this plant bloom and the ground was thawed enough for family members to bury those who had passed during the cold, frozen winter.

The common name shadbush refers to the season along the East Coast when the shad fish run upstream, which also coincides with the blooming of the serviceberry.

This is a prolific wild food and definitely worth getting to know.

STRAWBERRY
Fragaria spp.

Family: Rosaceae
Other names: Wild strawberry
Look-alikes: CInquefoil (leaves)
Related species: Woodland strawberry (*F. vesca*), Virginia strawberry (*F. virginiana*)

Description
More than twenty species worldwide, with many more subspecies and cultivars. All look similar, with some differences in leaf and fruit size and variations in geographic range and habitat.

Leaves sometimes look like a single leaf with three deep, sharply serrated lobes. Other leaves are more like three separate leaflets coming out of a single point at the tip of the leaf stalk. Leaves grow up to about 4" long and are always toothed, sometimes waxy, sometimes not. Leaves turn colorful in fall and make a pretty red-hued ground cover. Leaves are somewhat similar to the leaves of cinquefoil but almost always less accordion pleated.

Flower petals are short and fat (rounded with a slight tip) and white. Inflorescence is clustered, a few together, with a prominent a bright-yellow center.

Edible and Useful Plants

> ### RECIPE
> **Fresh Fruit Muesli**
>
> Combine 1 cup organic oats, ¾ cup rye flakes, ¼ cup dried cherries or raisins (unsweetened, unsulfured), ¼ cup pumpkin seeds, ¼ cup crushed or slivered almonds, ⅛ cup sunflower seeds, and ⅛ cup flaxseeds. Serve in a bowl with homemade almond milk or vanilla coconut yogurt, and top with plenty of fresh wild strawberries.
>
> **Variation:** Add additional fresh fruit such as blueberries, plums, and apples.

Strawberry plants are low-growing (up to 1' high) and spreading. They produce a basal rosette of coarsely serrated, three-lobed leaves. Plants send out long runners that set root in the soil. Strawberries are not really fruits but are commonly referred to as a fruit or berry. The berries are bright red and much smaller than store-bought varieties, usually less than 1". Achenes housing the seeds dot the outside of the berries. Depending on the species, achenes might protrude or be deeply seated in the red flesh.

Range and Habitat
Found throughout the entire continent. Large spreading mats on forest floors and in dappled shade, especially where somewhat moist.

Comments
Wild strawberries are small and often sparse. It's usually irresistible to eat the few that you can find straight from the plant. If some do manage to make it home with you, they make a great addition to desserts (decorate a cake or cookie platter), cereal, drinks, and salads.

Edible and Useful Plants   263

THIMBLEBERRY
Rubus parviflorus

Family: Rosaceae
Other names: Western thimbleberry, salmonberry, white-flowering raspberry, western thimble raspberry
Related species: Boulder raspberry (*Oreobatus deliciosus*), delicious raspberry (*R. deliciosus*)
Look-alikes: Baneberries (**poisonous**), raspberries (berries), currants (leaves), Rocky Mountain maple (leaves)
WARNING: The leaves of thimbleberry look very similar to those of the poisonous baneberry. The berries, however, are totally different between the two.

Description
This little-known native perennial shrub has some of the best-tasting wild berries in the Rockies. It has large leaves and is thornless. Fairly large, white flowers (rarely pink) have five distinct white petals with obvious fuzzy, yellow stamens forming a big, round, yellow center. Flowers are up to 2½" across.

Thimbleberry (*R. parviflorus*) has flowers in clusters of two to seven. Its leaves are about 4"–8" wide. Boulder raspberry (*Oreobatus deliciosus*) has solitary flowers. Its leaves are smaller, about 1"–2" wide.

RECIPE

Thimbleberry Tofu Pudding

Harvest ¼ cup thimbleberries. Cut 1 block firm tofu into quarters; place in a blender. Add 3 tablespoons honey and a dash of vanilla. Add 1/8 teaspoon cinnamon. Add 3 tablespoons water (more or less, depending on how much liquid your blender needs to work correctly). Blend until very smooth and creamy. Add the thimbleberries and blend in briefly.

Pour pudding into a bowl, using a soft spatula to scrape the sides of the blender. Serve immediately at room temperature, or chill in refrigerator for 2 hours or more and serve cold. Save a few berries to put on top as garnish.

Thimbleberry leaves are palmate, similar to a maple leaf, somewhat fuzzy and soft, with five shallow lobes. The largest leaves are about the size of a hand with outstretched fingers. The lobes can be pointed. Boulder raspberry leaves are more rounded and not fuzzy, and they are somewhat waxy looking in comparison. They look similar to the leaves of the wax currant but larger.

Edible and Useful Plants

Thimbleberry grows 1'–8' tall. Each spindly, scraggly cane lives 2 to 3 years. First-year primocanes do not produce fruit. After that, they may become branched and productive, sometimes reaching 8' tall. Greenish twigs have fine hairs. Reseeds in disturbed areas and can form dense thickets via its underground rhizomes.

Numerous drupelets form a berrylike cluster. "Berries" are similar to large raspberries but bigger and softer. Found alone (boulder raspberry) or in groups of two to seven (thimbleberry).

Range and Habitat
Alaska and British Columbia, across the western United States to New Mexico. Grows in woods, on open hillsides, in avalanche chutes, and along stream banks. Found in both moist and dry areas. From 3,500'–9,500' elevation.

Comments
Berries and young shoots are edible. Many accounts say that thimbleberries are dry and tasteless. When I finally found and tasted my first thimbleberry, I immediately thought I had eaten the wrong plant because that one was so incredibly soft, juicy, and sweet. Young shoots can be eaten, and leaves are used for medicinal purposes. Can be cultivated by transplanting rhizomes.

The best way to eat these is raw, straight from the branch. Take care picking, as they are quite mushy. If saving for later, use a sturdy container (not a bag) to harvest and transport the berries.

RUBIACEAE / MADDER FAMILY

NORTHERN BEDSTRAW
Galium boreale

Family: Rubiaceae
Other names: Cleavers, *G. septentrionale*
Related species: Sweet-scented bedstraw (*G. triflorum*), three-petal bedstraw (*G. trifidum*), cleavers (*G. aparine*)

Description
This delicate-looking native perennial is recognizable by its distinct, narrow leaves (lanceolate or linear) that form along the smooth stalk in whorls of four. Leaves are up to about 2" long, with some whorls of smaller leaves. Stalks grow 1'–3' tall. Other species of bedstraw have more rounded, wider leaves.

Showy, dense clusters of tiny whitish flowers bloom in spring. Northern bedstraw reproduces by seed and creeping rhizomatic root systems. Seed clusters look almost like flowers from a slight distance. They are rounded pink-to-cream capsules.

> ### RECIPE
>
> **Baked Cleaver Lasagna (Gluten-Free)**
>
> Preheat oven to 400°F.
>
> Cover the bottom of a heavy baking dish with 2 tablespoons olive oil.
>
> Layer the following ingredients in alternating layers: thinly sliced yellow summer squash and zucchini, sliced mushrooms, ground beef, bedstraw leaves, mozzarella cheese, fresh tomatoes, fresh garlic, oregano, rosemary, and thyme.
>
> Begin layering, starting with the zucchini or summer squash. Arrange in a layer, as you would lasagna noodles on the bottom of the baking dish.
>
> Next, add the meat and sprinkle on a layer of bedstraw leaves, garlic, tomato, cheese, and herbs. Continue layering until pan is full, ending with a top layer of cheese.
>
> Bake, covered, for 30 minutes. Rotate pan and remove lid. Continue cooking 15 to 25 minutes more, until bubbling and cheese begins to brown.

G. aparine is a very common species. It is a naturalized nonnative annual, whereas most bedstraws are perennial.

Range and Habitat

From Alaska to New Mexico in foothills, on montane slopes, and in moist meadows or woods. Also found west to California and across the northeastern United States and all of Canada.

Comments

Seeds and roots can be roasted and used similarly to coffee. Bedstraw leaves are a potherb and can be added to any soup or stew for extra nutrition. Leaves are small, so it does take some time to collect a sufficient amount.

The leaves of some species can be unpleasant raw. Their hairiness can upset the mouth and throat. Boiled, they are perfectly enjoyable.

RECIPE

Fresh Trout with Bedstraw Garnish

In a bowl, combine ½ cup coarse cornmeal, 1 tablespoon large-grain sea salt, 1 tablespoon fresh-ground black pepper, and 1 teaspoon cayenne powder. Mix well to combine.

Dunk 4 medium-size, moist trout fillets, one at a time, into the cornmeal mixture. Rotate bowl, and cover all sides of the fillet with as much cornmeal as possible. Use your hands to gently press cornmeal mixture into fish.

In a small saucepan, bring 3 cups water to a boil. Add 1 cup bedstraw leaves. Boil for 5 minutes and drain. Toss leaves in a bowl with the juice of 1 lemon and set aside.

Heat 4 tablespoons coconut or olive oil in a skillet to medium-high heat. When a drop of water sizzles on the oil, add the fish. Cook until just browned on bottom; flip them, and brown the other side. Remove from heat and place on plates. Garnish with the bedstraw leaves, gently laying the leaves out in a crisscross pattern across the top of the fish. Garnish with a lemon wedge. Serve atop a bed of mixed greens.

SCROPHULARIACEAE / FIGWORT FAMILY

GREAT MULLEIN
Verbascum thapsus

Family: Scrophulariaceae
Other names: Common mullein, lungwort, velvet dock, velvet plant, punchon, *gordolobo*, candlewick
Look-alikes: Evening primrose (*Oenothera biennis*)
WARNING: The **seeds are toxic** and should not be eaten. Do not confuse flower buds with seeds. Seeds are also toxic to fish, so they should not be thrown into waterways. The hairs on the leaves are rubefacient, or irritating to the skin.

Tea made with the flowers can be a mild sedative, so experiment with small amounts first. When making tea, be sure to strain it well to remove the fine hairs, which can cause significant irritation. The leaves contain rotenone (an insecticide) and coumarin (prevents blood clotting) but are not generally problematic when used in normal quantities.

> **RECIPE**
>
> **Mullein Leaf Tea**
>
> Great mullein leaves are often used to support lung health. Gather leaves from the basal rosette or stalk. Can be dried and stored. Bring water to a boil. Steep a few leaves and sip for a delicious wild tea to support the lungs.
> **Variation:** Add fresh or dried rose hips and wild mint.

Description
This nonnative Eurasian immigrant is a biennial with thick, soft, rounded ovular leaves. In the first year, the plant creates a basal rosette of large, thick, gray-green, felt-covered, rounded leaves. Leaves are covered in dense, soft hairs, giving it the look and feel of velvet, similar to lamb's-ear.

In the second year, the plant sends up a large, straight stalk, sometimes branched a few times, that produces a dense cone of yellow flowers along the stalk. Flowers, flower buds, and seedpods are packed together and densely cover the stalk. Flowers open and go to seed at different times, even on the same stalk.

Mullein stalks are thick, about 2" in diameter, and grow 1"–8" tall. Note that in some climates and conditions, mullein can be an annual or a short-lived perennial.

Range and Habitat
From Alaska south across the entire United States, including Hawaii. Mullein grows in sunny, disturbed areas; along roadsides; and in vacant lots and open fields. Found from sea level to tree line.

Comments
Many people consider great mullein medicinal only. Others consider it edible.

> FORAGER NOTE: **A word of caution!** Do not eat the seeds, which are poisonous. The long mullein stalk often hosts both flowers and seeds at once. Seeds can often be found right next to emerging buds. Be careful when picking flower buds. **Do not confuse flower buds with seedpods.** It is easy to do so. **The seeds are poisonous.** The flowers bloom and go to seed at different times on the same stalk, so confusion is very possible.

Flower buds and flowers are edible raw, cooked, or dried. Flower buds make a wonderful trail snack. They are tender and sweet.

Leaves are generally washed, dried, and used as medicinal tea.

The fine hairs on all parts of the plant cause irritation in the mouth and throat and need to be strained out through a cheesecloth or fine sieve.

My favorite way to use great mullein is as a blush. Harvest one leaf and rub gently on the apple of the cheek for a healthy glow.

Mullein is also called lungwort because it is used as a medicinal herb for hay fever and asthma, to reduce inflammation in lungs, and for ear infections.

The soft, thick, velvety leaves can be used as liners for shoes, toilet paper (directionality is everything), or menstrual pads. Flowers can be made into a hair dye. Mullein can be used as a poultice to reduce swelling, to help heal sprains, and as an expectorant. The plant also contains vitamins B_2, B_5, B_{12}, and D; choline; hesperidin; para-aminobenzoic acid; magnesium; and sulfur.

RECIPE

Great Mullein Torch

Great mullein can be dipped in suet or wax and used as a torch. Melt candle wax and dip the mullein stalk several times to create multiple wax layers. Let dry in between layers. Take care with open flame and **never** leave unattended or burn where there is risk of starting a fire.

TYPHACEAE / CATTAIL FAMILY

CATTAIL
Typha latifolia

Family: Typhaceae
Other names: Common cattail, broadleaf cattail
Look-alikes: Young shoots may resemble poisonous death camas, western blue flag, or other nonedible iris family members.
Related species: Narrowleaf cattail (*T. angustifolia*), which is very similar, but a small gap of less than 1"–4" separates the female and male portions of the flower along the flower stalk. The narrow-leaved variety grows slightly shorter, 3½'–5' tall, and the leaves are about ¼" wide.
WARNING: Some sources report that cattail should be avoided during pregnancy, others do not. Cattails are filter plants, and if they are growing in contaminated water or where there is contaminated runoff, they should **NOT** be consumed by anyone.

Description

Tall, narrow-leaved perennials growing in dense stands. Easy to identify by the large, dark brown, "hot dog on a stick" appearance of the flowers. Tall flower stalks are interspersed among the tall, straight green leaves. Plants reach 3'–9' in height. Erect, stiff leaves are tall and loosely surround the stalk; they are less than 1" wide.

Flowers are deceivingly un-flowerlike, since the flowers do not have petals. Female flowers grow in dense bunches toward the top of the stalks and form a dense brown tube or spike. Just above the female portion, at the tip of the stalk, are the male flowers, which are yellow and form a much thinner tube up to the pointed tip of the stalk. They are short-lived. It is the female portion of the flower that gives cattails their distinct brown tubular appearance.

The flower spikes are less noticeable when they are young and still green. As they age, they become brown and, eventually, they float away similar to the cotton-like seedpods of dandelion or thistle.

Range and Habitat

Cattails grow in swamps, marshes, and calm, moist riparian zones from Canada to New Mexico. They are widespread in moist areas throughout the United States but not generally found in the higher elevations of the Rocky Mountains. They are often found in roadside drainages and pristine riparian areas throughout the region.

Comments

Cattails are highly edible and nutritious. The young flower spikes (green buds before they turn brown), corms (swollen tuber, bulb), rhizomes (thick lateral roots), lower portion of the leaves, peeled stalk, pollen, and seeds are all edible.

RECIPE

Pollen Pancakes

Harvest yellow pollen from the male flowers by shaking them into a container or bag. There is no need to cut down the cattail to do this. There is a lot of pollen, so it is fairly simple to gather enough for a batch of pancakes. Substitute ¼ of the flour or pancake mix with pollen and proceed as usual. Don't over-sweeten with maple syrup or you will drown out the flavor of the pollen.

WARNING: People with allergies to other pollens should be extra careful.

To eat the tender lower portion of the leaves, grasp the leaf at the base and pull it free. Scrape the flesh off, chop and add to salad.

The young shoots are a delicious and tender vegetable and can be used similar to asparagus or bamboo. Peel the outer layers and then boil, steam, or sauté the inner part of the stalk or eat raw.

Young green flower spikes can be grilled or boiled much like corn on the cob. Pollen can be shaken from the male flower spikes and used as a nutritious addition to pancakes and baked goods or as a roux to thicken stew.

The roots are also edible but quite fibrous. They are prepared by forcibly separating the fiber strands from the starchy root. Since cattails are perennials, harvesting the root will kill the plant, so be sure to harvest roots only from robust stands where there will be plenty of cattails left.

Old-timers dipped the flower stalk in bear fat to make lanterns. It has also been said that you can burn the brown part off as a way of winnowing the seeds, which can also be used as food. The stiff leaves of cattails can be used to weave baskets, floor mats, seats, and water-storage containers. The fluffy down from the seedpods can be used as stuffing for pillows, mattresses, and, supposedly, even for life jackets.

If you experiment with the life jacket or safety flotation idea, please do not rely on it for lifesaving purposes. You will need a properly classified personal floatation device for anything beyond experimentation.

RECIPE

Grilled Cattail on the Cob

Harvest young, green female flower spikes. Place on grill over medium heat. Rotate so that no side becomes burned. Cook until the spikes just begin to brown; remove from heat. Eat like corn on the cob, plain or with butter and salt.

Variation: Instead of butter and salt, douse with fresh-squeezed lime juice and homemade mayonnaise, and top with roasted chili powder.

URTICACEAE / NETTLE FAMILY

STINGING NETTLE
Urtica dioica

Family: Urticaceae
Other names: California nettle, slender nettle, tall nettle
Look-alikes: All members of the mint family also have square stems.
WARNING: As the name suggests, this plant stings! It can cause rash, hives, and allergic reactions. Harvest it while wearing gloves and a long-sleeved shirt. It can cause uterine contractions, so pregnant women should avoid contact. Some sources warn that the plant should not be consumed after it begins to seed because chemistry changes make it unsafe to consume. Some sources also warn that older leaves can irritate the kidneys, but there is debate about this. Also, it is said to interfere with allopathic drugs used for ailments such as diabetes and hypertension. Be sure to consult with your physician if this applies to you.

Description

This herbaceous perennial has long, thin, deeply serrated, hairy, ovular or lance-shaped leaves with pointed tips. Leaves are simple and opposite and often hang downward. The undersides of the leaves have hollow, needlelike hairs. The hairs contain a toxin that stings and can cause a rash and hives, which usually dissipate within a day.

Thick, erect, square stalks are hairy and 1½"–10' tall, usually 3'–4' high. Stalks are up to 1" thick.

Small whitish, greenish, or pinkish flowers have sepals but no petals and cluster in spikes along the stalk; clusters are about 4" long. The plant becomes thickly laden with green, narrow, drooping clusters of seedpods that hang off the stem.

Range and Habitat

From Alaska to Florida. Moist fields, moist disturbed ground, open woods, and rocky areas in the plains, foothills, and montane zones. Grows in dense, sporadic patches.

RECIPE

Creamy Nettle Soup with Bacon

Bring 8 cups vegetable broth to a boil. Add 2 carrots (sliced) and 1 medium-size potato (diced). Simmer for 12 minutes.

Meanwhile, fry 6 strips of bacon until cooked but not crisp. Remove from heat; set aside.

To the broth mixture, add 1½ cups chopped young shoots and leaves (or older leaves) of stinging nettle. Simmer for 12 minutes. Put broth-nettle mixture in blender and process until creamed. Return to pot.

In a small saucepan, heat 1 teaspoon lard or olive oil to medium. Sauté 2 tablespoons all-purpose flour, constantly stirring for 2 minutes. Remove from heat and stir into soup until blended and soup looks creamier. Add crumbled bacon.

Simmer very low so that the flavors combine, about 15 to 20 minutes. Serve with homemade crostini, black sea salt, and fresh-ground pepper.

To make crostini: Cut bread into 2" squares. Lay squares out on a cookie sheet and drizzle with olive oil. Sprinkle with salt, pepper, and paprika. Bake in oven at 275°F until dry and just crisp, about 7 minutes.

Comments

Young shoots, young leaves, and the root are edible. Cooking or drying eliminates the sting of the toxin and actually turns stinging nettle into a delicious and versatile vegetable. The young shoots and leaves are particularly tender before the plant flowers, but leaves can also be collected after first flowering. Can be cooked like any vegetable.

Some sources say the roots can be eaten cooked similar to a potato. Tea can also be made and drunk fresh brewed or fermented.

Fibers can be used to make fine textiles or rope. Roots, seeds, and greens are used for medicinal purposes. Use leaves to make a cleansing tea along with other greens or dried as a nutritional supplement.

On a rock-climbing trip in Wyoming, I pointed out a patch of stinging nettle. My good friend Ted, always up for adventure, immediately helped confirm the identification by thrashing his thick, bare arm about in the nettle patch. The red welts, itching, and hives that persisted until the end of the day proved that this plant was indeed stinging nettle.

VIOLACEAE / VIOLET FAMILY

VIOLET
Viola spp.

Family: Violaceae
Related species: Western dog violet/early blue violets (*V. adunca*), Canada violet (*V. canadensis*), round-leaved yellow violet (*V. orbiculata*), yellow montane violet/yellow prairie violet (*V. nuttallii*)

Description

Violets are low-growing and can be annual or perennial. Usually, the leaves are in a basal cluster, but they can be alternate along the short stalks. The classic leaf shape is heart shaped, but some varieties are more lanceolate or linear.

Showy flowers are purple, white, or yellow, depending on species. They have five petals, with two matching petals pointing upward, two petals that point one to each side or somewhat downward, and a single bottom petal that is somewhat bulged and points down. The single petal is often notable by the purple, brown, blackish, or pink stripes emanating out and down from the flower's center. These

> ### RECIPE
>
> **Fresh Violet Salad**
>
> Harvest fresh leaves and flowers in spring or summer. If you can find several violet varieties (yellow and purple), that is even better. Toss delicate salad greens, such as butter lettuce, and violet greens with a light, crisp olive oil and a good plum vinegar. Lightly salt and pepper. Top with violet flowers, making a pretty pattern of bright yellow and purple. This is an astonishingly gorgeous way to eat your greens.

markings might also extend from the side petals. The flower head is usually open and showy.

Range and Habitat
Throughout the United States, but specific ranges vary depending on species. Found in moist, dappled shade, dry meadows, gardens, and, for some species, up to subalpine zone.

Comments
These pretty flowers are considered a weed by some, an ornamental by others, and a special treat by foragers. All violets are edible. Leaves and flowers can be eaten raw or cooked or drunk as a tea. Flowers can be made into candy and wine.

HONEYBEES

This chapter is for fun only. Please do not harvest wild honey. Bee populations are in massive collapse due to habitat destruction and toxic agriculture. Enjoy the knowledge that wild honey is buzzing around you, but leave it for the bees, and lobby your politicians to fix this crisis.

HONEYBEES
Apis mellifera

Family: Apidae
Other names: European honeybee, western honeybee
WARNING: Honeybees do sting, although much less often than wasps. Honeybees are usually very docile when encountered away from the hive and will not sting unless threatened. However, they will sting if one gets too close to a hive or tries to get to the honey.

Edible and Useful Plants   283

A sting from one bee is usually not too bad, but stings by many bees can be quite painful. Some people are **deathly allergic** to bee stings.

Pay very close attention to what happens after you or someone else is stung. Most people will get a small welt, and the pain will subside after several minutes. If a person is allergic to bee stings, the reaction can be much worse, even deadly. If the welt grows; hives appear anywhere on the body; the eyes, throat, face, or any part of the body begin to swell; or the person has trouble breathing or asthma-like symptoms develop, **seek medical attention immediately**. Anaphylaxis and death can result.

Description

Honeybees have three main body parts—head, thorax, and abdomen—and are about ½" long. Honeybees typically have alternating orange-yellow and black stripes. Honeybee midsections are rounded, and they are not constricted like those of wasps. They are densely hairy, less so on the abdomen.

Generally, there are three different types of bees within a colony. The females consist of the worker bees and, usually, one active queen. The drones are the males. The queen is noticeably larger than the others and typically can be distinguished by her larger size and lack of stripes. The queen's sole purpose is to lay eggs and be fed and cared for by the worker bees. The drones' only purpose is to mate with the queen.

> FORAGER NOTE: When we think of foraging, we often forget that honey exists in the wild. What a treat. Bees are struggling enough as it is with the extensive use of pesticides and herbicides and because of so much habitat loss. Please do not harvest wild honey.

Worker bees do most of the work in the colony, including cleaning the hive; feeding and cleaning larvae; tending to the queen; guarding and patrolling the hive; heating or cooling the hive; and foraging for pollen, nectar, honeydew,

RECIPE

Hot Honey Sundae

Perfect to share in one big bowl with your honey. Choose your favorite version of vanilla ice cream (organic cow's milk ice cream; icy sorbet; or a vegan, nondairy option). In a small saucepan over very low heat, heat some honey until just warm. Spoon honey on top of two scoops of ice cream. Top generously with crushed peanuts.

water, and tree sap. It is from the collection of these things that the bees produce honey. Honey is produced to feed the colony and to get the colony through the winter, when the entire colony hibernates in the hive.

Comments

In addition to the potential for getting stung, wild honey collecting is **not advised** due to the massive decline in honeybee populations around the world. At this time in history, it is essential that we all take the utmost care to protect the bee populations. Given the extreme importance of honeybees to the pollination of flowering plants and the stress they are under due to widespread pesticide use, genetically modified organism (GMO) monocropping, habitat destruction, and harmful industrial beekeeping practices, bees in the wild should be revered and left to continue their natural life cycle. Taking their honey places unnecessary stress on them. **Please, do not harvest wild honey.**

Honeybees are one of nature's most remarkable creations, both because they produce delicious honey and also because they have a very complex social order.

There are many other types of bees, ranging from a tiny bee barely millimeters long to a very large bee that can be 1½" long. There are also many species of wasps that have features similar to bees. But only the honeybee of the genus *Apis*, of which there are eight species, actually produces honey.

INDEX

*Recipes

Achillea millefolium, 109–11
Aconitum spp., 8–10
*Acorn Porridge, 184
Actaea rubra, 2–3
Agastache urticifolia, 191–93
alfalfa, 167–69
*Alfalfa Supplement, Dried, 168
Allium cernuum, 197–99
alpine sorrel, 233–34
alpine zone, xxxii
amaranth
    common, 23–24
    creeping, 25–27
amaranth family, 23–27
*Amaranth Leaf Quesadilla, 24
Amaranthus blitoides, 25–27
Amaranthus retroflexus, 23–24
Amelanchier alnifolia, 257–60
American plum, 246–48
Anaphalis margaritacea, 82–83
*Ants on a Log, 37
Apis mellifera, 283–85
apple, 240–42
*Apple-Mallow Fruit Salad, 203
*Apples, Baked, 242
*Apples, Dried, 237
*Apples, Frozen, 242
Arctium spp., 61–63
Arctostaphylos uva-ursi, 161–63
arnica, heart-leaved, 53–56
Arnica cordifolia, 53–56
*Arnica Oil, 55
Artemesia frigida, 89–90
Artemesia ludoviciana, 91–93
Artemisia tridentata, 57–60
Asclepias speciosa, 46–49
Asparagus, 50–52
    *Grilled, 51
asparagus family, 50–52
Asparagus officinalis, 50–52
aster or daisy family, 53–111

*Backyard Sun Tea, 84
*Bacon Fried Young Greens (dandelion), 74
*Baked Apples, 242

*Baked Cleaver Lasagna (Gluten-Free), 268
baneberry, 2–3
barberry family, 112–15
bedstraw, northern, 267–69
*Bedstraw Garnish, Fresh Trout with, 269
beech family, 182–84
bellflower family, 128–29
big sagebrush, 57–60
bistort, 226–28
*Bistort Root, Roasted, 228
Bistorta bistortoides, 226–28
bittercress, heart-leaved, 118–20
*Bittercress Easy Omelet, 120
*Bittercress Salad, 119
*Blite Salad, 138
blue spruce, 212–14
*Blue Spruce Beer, 213
bluebell, mountain, 116–17
*Bluebell and Chickpea Fresh Summer Salad, 117
blueberry, 164–66
*Boiled Root and Leaves with Winter Stew (evening primrose), 206
*Boiled Stem (nodding thistle), 108
borage family, 116–17
buckwheat family, 226–37
buffaloberry, Canada, 149–50
bull thistle, 101–3
burdock, 61–63

cactus family, 125–27
Campanula rotundifolia, 128–29
Canada buffaloberry, 149–50
Capsella bursa-pastoris, 123–24
caraway, 33–35
    *Seed with Grilled Beets and Turnips, 35
Cardamine cordifolia, 118–20
Carduus nutans, 107–8
*Carrot and Parsley Salad, 45
carrot family, 33–45
Carum spp., 33–35
cattail, 273–75
    *Grilled, on the Cob, 275
cattail family, 273–75
century plant family, 20–22

287

*Ceremonial Burning (big sagebrush), 59
chamomile, 64–66
*Chamomile Eye Pillow, 66
*Chamomile Mint Tea, 65
Chenopodium album, 133–35
Chenopodium capitatum, 136–38
chicory, 67–69
*Chicory Root Coffee, 69
chokecherry, 243–45
*Chokecherry Jelly, 244
Cichorium intybus, 67–69
Cicuta spp., 14–15
Cirsium arvense, 104–6
Cirsium vulgare, 101–3
*Classic Dairy-Free Basil Pesto, 220
clover, red, 170–72
clover, sweet, 173–75
 *Garnish, with Grilled Chicken, 175
clover, white, 176–78
*Clover and Rice Breakfast, Dried, 178
*Clover Flowers, Sautéed, with Candied Pecans, 171
*Clover Salad, Fresh, 170
*Clover Tea, Red, 172
*Clover Tea, Sweet, 174
*Cold *Agastache* Gazpacho, 192
common
 amaranth, 23–24
 juniper, 143–45
 plantain, 224–25
 rabbitbrush, 86–88
 sunflower, 98–100
Conium maculatum, 11–13
cow parsnip, 36–68
 *Sautéed Stalks and Leaf Stems, 38
*Creamy Nettle Soup with Bacon, 277
creeping amaranth, 25–27
*Creeping Amaranth Macaroni and Cheese, 26
creeping thistle, 104–6
curly dock, 229–32
currant, 185–88
currant, prickly, 189–90
currant family, 185–90
cutleaf coneflower, 70–71
cypress family, 143–48

dandelion, 72–75
Daucus carota, 43–45
Delphinium spp., 6–7
dock, curly, 229–32

*Dock Leaf Nachos, 231
*Doggie Sachet, 77
Douglas fir, 215–17
*Dried Alfalfa Supplement, 168
*Dried Apples, 242
*Dried Clover and Rice Breakfast, 178
*Dried Serviceberries, 258
*Dried Wild Plums, 247

Elaeagnus angustifolia, 151–53
elderberry, 130–32
*Elderberry Cold and Flu Syrup, 132
Ephedra spp., 154–55
Epilobium angustifolium, 208–10
Equisetum arvense, 156–58
Equisetum hyemale, 159–60
Ericameria spp., 86–88
Erigeron spp., 76–78
evening primrose, 205–7
evening primrose family, 205–10

figwort family, 270–72
fireweed, 208–10
flax, western blue, 200–201
*Flax and Nut Crust, Raw Pumpkin Pie with, 201
flax family, 200–201
fleabane, 76–78
*Flower Stalk Tea (sage), 90
*Foot Soak (big sagebrush), 59
foothills, iii
forget-me-not family, 116–17
Fragaria spp., 261–63
*French Lentils and Alfalfa Flower Heads, 169
*Fresh Clover Salad, 170
*Fresh Fruit Muesli, 262
*Fresh Greens Salad (dandelion), 73
*Fresh Mint Sorbet, 195
*Fresh Trout with Bedstraw Garnish, 269
*Fresh Violet Salad, 281
*Fried Cambium (ponderosa pine), 223
*Fried Tofu with Sweet-Spicy Rose Hip Sauce, 255
fringed sage, 89–90
*Frozen Apples, 242
*Fruit Muesli, Fresh, 262
*Fruit Salad with Fresh Mint, 196

Galium boreale, 267–69
Gambel oak, 182–84

*Garlic Fava Beans and Chicory Greens, 68
giant hyssop, 191–93
Glycyrrhiza lepidota, 179–81
*Gobo, 62
golden banner, 4–5
goldenrod, 79–81
*Goldenrod Seed Chicken Soup, 80
*Good Old-Fashioned Sausage, Onion, and Purslane, 239
goosefoot family, 133–38
great mullein, 270–72
*Great Mullein Torch, 272
*Green Juice, 24
*Greens Salad, Fresh (dandelion), 73
*Grilled Asparagus, 51
*Grilled Beets and Turnips with Wild Caraway Seed, 35
*Grilled Cattail on the Cob, 275
*Grilled Chicken with Yellow Sweet Clover Garnish, 175
*Grilled Nopal Salsa, 126
*Grilled Plum and Goat Cheese Salad, 248

harebell, 128–29
*Harebell Leaf and Flower Salad, 129
heart-leaved arnica, 53–56
heart-leaved bittercress, 118–20
*Hearty Wild Currant Pancakes, 186
*Hearty Wild Game Stew, 45
heath family, 161–66
Helianthus annuus, 98–100
hemlock
    poison, 11–13
    water, 14–15
Heracleum maximum, 36–38
*Holiday Necklaces, 144
*Homemade Toothpaste, 160
*Honey, Raw, and Osha Root (for colds and flu), 41
honeybees, 283–85
honeysuckle family, 130–32
horsetail, 156–58
horsetail family, 156–60
*Horsetail Tea, 158
*Hot Cereal and Fresh Raspberries, 252
*Hot Honey Sundae, 284
huckleberry, mountain, 164–66
hyssop, giant, 191–93

*Indian Ice Cream, 150

juniper
    common, 143–45
    Rocky Mountain, 146–48
*Juniper Berries, Crushed, with Simple Lamb or Venison Stew, 148
*Juniper Berries, Raw, 148
Juniperus communis, 143–45
Juniperus scopulorum, 146–48

*Kasha with Creeping Amaranth, 26
kinnikinnick, 161–63

lamb's-quarter, 133–35
larkspur, 6–7
*Lemon Pudding with Fresh Wild Currants, 187
*Lemonade Cocktail with Fresh Juniper, 145
licorice, wild, 179–81
*Lightly Boiled Flower Buds (bull thistle), 103
Ligusticum porteri, 39–42
lily family, 197–99
Linum lewisii, 200–201

madder family, 267–69
*Mahonia and Watermelon Juice, 114
Mahonia repens, 112–15
mallow, 202–4
mallow family, 202–4
*Mallow Gumbo, 204
*Mallow Wheels, Raw, 204
Malus domestica, 240–42
Malva neglecta, 202–4
*Marrow and Ground Cones (Douglas fir), 216
Matricaria discoidea, 84–85
Matricaria recutita (*Matricaria chamomilla*), 64–66
Medicago sativa, 167–69
*Medicinal Tea (cutleaf coneflower), 71
Melilotus officinalis, 173–75
Mentha spp., 194–96
Mertensia ciliata, 116–17
Mertensia spp., 116–17
milkweed, 46–49
milkweed family, 46–49
mint, 194–96
mint family, 191–96
*Mint Sorbet, Fresh, 195
*Mint Tea, 196
monkshood, 8–10
montane zone, xxxiii

Index   **289**

Mormon tea, 154–55
*Mormon Tea, 155
Mormon tea family, 154–55
mountain bluebell, 116–17
mountain huckleberry, 164–66
mullein, great, 270–72
*Mullein Flower Tea, 271
*Mullein Torch, Great, 272
mustard family, 118–24

narrowleaf yucca, 20–22
*Needle Tea (Douglas fir), 217
nettle, stinging, 276–79
nettle family, 276–79
nodding onion, 197–99
nodding thistle, 107–8
*Nopal Fruit Salad, 127
*Nopal Salsa, Grilled, 126
northern bedstraw, 267–69

Oenothera biennis, 205–7
*Old-Fashioned Raspberry Pie, 250
oleaster family, 149–53
olive, Russian, 151–53
onion, nodding, 197–99
*Onion with Brown Rice, and Sautéed Shrimp, 198
Opuntia spp., 125–27
Oregon grape, 112–15
*Oregon Grape Flower Popsicles, 115
osha, 39–42
*Osha Root Tea, 42
oxalates, xxii
Oxyria digyna, 233–34

pea family, 167–81
*Peach and Clover Salad, 177
pearly everlasting, 82–83
    *Simple Sautéed Greens, 83
*Pemmican, 162
pennycress, 121–22
*Pennycress Coconut Soup with Bok Choy and Tofu, 122
Picea pungens, 212–14
pine, piñon, 218–20
pine, ponderosa, 221–23
pine family, 211–23
*Pine Nut Trail Mix, 219
pineapple weed, 84–85
Pinus edulis, 218–20

Pinus ponderosa, 221–23
Plantago major, 224–25
plantain, common, 224–25
plantain family, 224–25
plum, American, 246–48
    *Grilled, and Goat Cheese Salad, 248
poison hemlock, 11–13
poison ivy, western, 16–17
*Polish Sorrel Soup, 236
*Pollen Pancakes, 274
*Ponderosa Needle Tea, 222
ponderosa pine, 221–23
Portulaca oleracea, 238–39
*Potato Crusted Onion Frittata, 199
*Potherb, 82
*Poultice for Insect Bites, 225
*Poultice to Stop Bleeding (yarrow), 111
prickly currant, 189–90
prickly pear, 125–27
*Prickly Pear Juice, 127
primrose, evening, 205–10
Prunus americana, 246–48
Prunus virginiana, 243–45
Pseudotsuga menziesii, 215–17
purslane, 238–39
purslane family, 238–39

Queen Anne's lace, 43–45
Quercus gambelii, 182–84

rabbitbrush, common, 86–88
*Rabbitbrush and Artemisia Soak, 88
raspberry, 249–52
*Raw Honey and Osha Root (for colds and flu), 41
*Raw Juniper Berries, 148
*Raw Mallow Wheels, 204
*Raw Pumpkin Pie with Flax and Nut Crust, 201
red clover, 170–72
*Red Clover Tea, 172
Rhus glabra, 28–30
Rhus trilobata, 31–32
Ribes lacustre, 189–90
Ribes spp., 185–88
*Rice Bowl with Wild Greens, 134
riparian areas, xxxiii
*Roasted Bistort Root, 228
*Roasted Wild Licorice Root, 181
Rocky Mountain juniper, 146–48

*Romantic Vanilla Cake with Harebell
   Blossom, 129
Rosa spp., 253–56
rose, wild, 253–56
rose family, 240–66
roseroot, 139–40
Rubus parviflorus, 264–66
Rubus spp., 249–52
Rudbeckia laciniata, 70–71
Rumex acetosella, 235–37
Rumex crispus, 229–32
Russian olive, 151–53
*Russian Olive Jelly, 152

sage
   fringed, 89–90
   white, 91–93
sagebrush, big, 57–60
salsify, 94–97
*Salsify Leaves Pasta Primavera, 96
*Salty Fish Salad, 142
Sambucus nigra, 130–32
*Sautéed Clover Flowers with Candied
   Pecans, 171
*Sautéed Shrimp and Onion with Brown
   Rice, 198
*Sautéed Stalks and Leaf Stems (cow
   parsnip), 38
scouring rush, 159–60
Sedum integrifolium, 139–40
Sedum lanceolatum, 141–42
Sedum rhodanthum, 139–40
*Serviceberries, Dried, 258
serviceberry, 257–60
sheep sorrel, 235–37
Shepherdia canadensis, 149–50
shepherd's purse, 123–24
*Simple Compote (serviceberry), 259
*Simple Lamb or Venison Stew with Crushed
   Juniper Berries, 148
*Simple Sautéed Greens (pearly
   everlasting), 83
skunkbush, 31–32
smooth sumac, 28–30
*Smoothie Topped with Primrose Seeds, 207
*Smudge, 93
Solidago spp., 79–81
sorrel, alpine, 233–34
sorrel, sheep, 235–37
spear thistle, 101–3

spruce, blue, 212–14
*Steamed Young Shoots (fireweed), 209
*Stewed Anise Plums and Peaches, 193
stinging nettle, 276–79
*Stir-Fry with Milkweed Flowers, 48
stonecrop, 141–42
stonecrop family, 139–42
strawberry, 261–63
strawberry blite, 136–38
*Stuffed Zucchini with Dock, 230
subalpine zone, xxxii
sumac, smooth, 28–30
sumac, three-leaved, 31–32
*Sumac Juice, 30
sumac family, 28–32
sunflower, common, 98–100
*Sunflower Seed Muesli, 100
sweet clover, 173–75
*Sweet Clover Tea, 174

*Taproot and Fresh Greens Stew, 102
Taraxacum officinale, 72–75
Thermopsis spp., 4–5
thimbleberry, 264–66
*Thimbleberry Tofu Pudding, 265
thistle
   bull, 101–3
   creeping, 104–6
   nodding, 107–8
   spear, 101–3
Thlaspi arvense, 121–22
three-leaved sumac, 31–32
*Three-Leaved Sumac Lemonade, 32
*Three-Leaved Sumac Tapioca Pudding, 32
Toxicodendron rydbergii, 16–17
Tragopogon dubius, 94–97
*Trail Chew (Morman tea), 155
*Trail Recipe (sheep sorrel), 237
*Trail Snack, 165
*Trifecta Salsify Sampler, 95
Trifolium pratense, 170–72
Trifolium repens, 176–78
Typha latifolia, 273–75

Urtica dioica, 276–79

Vaccinium myrtillus, 164–66
Verbascum thapsus, 270–72
Viola spp., 280–82
violet, 280–82

violet family, 180–82
*Violet Salad, Fresh, 281

*Warm Roseroot with Spinach, 140
water hemlock, 14–15
western blue flax, 200–201
western poison ivy, 16–17
white clover, 176–78
white sage, 91–93
*Wild Greens Wontons, 124
wild licorice, 179–81
*Wild Licorice Root, Roasted, 181

*Wild Plums, Dried, 247
wild rose, 253–56
*Wild Spinach Salad, 135

yarrow, 109–11
*Yarrow Tea, 111
yucca, narrowleaf, 20–22
*Yucca Flower Sauté, 21
Yucca glauca, 20–22
*Yucca Shampoo, 22

zones, xxxii–xxxiii

ABOUT THE AUTHOR

Liz Brown Morgan, MA, JD, FNTP, RWP, has explored the relationship between food, health, and sustainability from many angles.

As an anthropology student at Colgate University, Liz studied how ancient civilizations ate and how they interacted with their environments to hunt, fish, forage, prepare, and eat food. As a wilderness guide, Liz spent countless hours exploring the wildlands of the American Northeast and Rocky Mountains, becoming familiar with the plants and ecosystems and connecting people with them.

As an environmental lawyer, Liz fought to protect the wild places and explored the role of humans as protectors of the natural lands. As an eco-entrepreneur, Liz created a green online marketplace demonstrating that business has a vital role in normalizing sustainability and socially and environmentally responsible practices.

As a functional nutritionist, she dove into supporting people with serious and mysterious health conditions to become nourished and reclaim their health. Liz is currently the executive director of a membership-based nonprofit in her community.

Liz is a food culture transitionista dedicated to creating a food culture that restores the health of the people and the planet. She believes in collectively becoming the wise elders future generations need, and that means passing on a functioning food culture so people can thrive in good health on a living, vibrant planet.

Liz has spent over two decades exploring the rivers, mountains, and wild lands of Colorado and the American West. She always stops to smell the wild roses.

Connect with Liz at www.linkedin.com/in/liz-morgan-food-health-sustainability.